"十三五"江苏省高等学校重点教材（编号：2016-2-089）

流体力学基础（双语版）
Fundamentals of Fluid Mechanics

周剑锋　邵春雷　主编
顾伯勤　主审

　化学工业出版社

·北京·

内容简介

本书系"十三五"江苏省高等学校重点教材,本书内容涉及流体力学基本概念、基本理论和研究方法等几个方面,全书共9章,包括流体力学简介、流体静力学、流体运动学、流体流动的有限控制体分析、流体流动的微分分析、相似理论和量纲分析、管内流动、平面势流和绕物流动。与其他流体力学教材不同,本书是以介绍流体力学基础知识为主,内容简单。采用双栏格式,中英文对照,可提高学生英文阅读能力。为了加强学生对内容的理解,每章均配有习题。

本书可作为高等院校非力学专业本科生流体力学课程教材,也可供相关人员参考使用。

图书在版编目（CIP）数据

流体力学基础：双语版：英、汉/周剑锋,邵春雷主编.—北京：化学工业出版社,2017.11（2024.8重印）
ISBN 978-7-122-31067-5

Ⅰ.①流… Ⅱ.①周…②邵… Ⅲ.①流体力学-高等学校-教材-英、汉 Ⅳ.①O35

中国版本图书馆CIP数据核字（2017）第292489号

责任编辑：程树珍　丁文璇　　　　　　　　　装帧设计：关　飞
责任校对：吴　静

出版发行：化学工业出版社（北京市东城区青年湖南街13号　邮政编码100011）
印　　装：北京盛通数码印刷有限公司
787mm×1092mm　1/16　印张16½　字数410千字　2024年8月北京第1版第5次印刷

购书咨询：010-64518888　　　　　　　　　　售后服务：010-64518899
网　　址：http://www.cip.com.cn

凡购买本书,如有缺损质量问题,本社销售中心负责调换。

定　价：59.00元　　　　　　　　　　　　　　　　　　版权所有　违者必究

前言
Preface

本书是为高等院校非力学专业本科流体力学课程中英双语教学编写的。考虑到目前流体力学课程的教学课时数有限,所以本教材以介绍流体力学基础知识为主,内容浅显,易于理解。本书旨在介绍流体力学的基本概念、基本理论和解决流体力学问题的基本方法,为本科生后续课程的学习及未来从事的工作提供必要的流体力学基本知识,同时提高学生阅读相关英文科技文献的能力。

本书共9章,包括流体力学简介、流体静力学、流体运动学、流体流动的有限控制体分析、流体流动的微分分析、相似理论和量纲分析、管内流动、平面势流以及绕物流动。全书文字部分采用双栏格式排版,中英文对照。为便于学生学习和理解教材内容,每章都附有一定数量的例题和习题。

本书第1章至第5章由南京工业大学邵春雷编写,第6章至第9章由南京工业大学周剑锋编写,全书由南京工业大学顾伯勤主审。本书得到了江苏高校品牌专业建设工程项目(PPZY2015A022)的资助。

限于编者水平,书中不足之处在所难免,恳请读者批评指正。

编　者
2017年10月10日　于南京

目录
Contents

Chapter 1
Introduction of Fluid Mechanics / 1

   1.1 Brief History of Fluid Mechanics / 1
   1.2 Dimensions and Units / 4
      1.2.1 Dimensions / 4
      1.2.2 Units / 8
   1.3 Definition of a Fluid / 10
      1.3.1 Continuity Hypothesis / 10
      1.3.2 Density / 12
      1.3.3 Specific Weight / 13
      1.3.4 Specific Gravity / 14
   1.4 Fluid Properties / 14
      1.4.1 Compressibility / 14
      1.4.2 Surface Tension / 17
      1.4.3 Viscosity / 20

Chapter 2
Fluid Statics / 27

   2.1 Pressure at a Point / 27
   2.2 Basic Equation for Pressure Field / 28
   2.3 Pressure Variation in a Fluid at Rest / 31
      2.3.1 Incompressible Fluid / 31
      2.3.2 Compressible Fluid / 34
   2.4 Standard Atmosphere / 35
   2.5 Buoyancy and Stability / 37
      2.5.1 Archimedes' Principle / 37
      2.5.2 Stability / 40
   2.6 Measurement of Pressure / 42
   2.7 Manometry / 44

第 1 章
流体力学简介 / 1

   1.1 流体力学简史 / 1
   1.2 量纲和单位 / 4
      1.2.1 量纲 / 4
      1.2.2 单位 / 8
   1.3 流体的定义 / 10
      1.3.1 连续性假设 / 10
      1.3.2 密度 / 12
      1.3.3 重度 / 13
      1.3.4 比重 / 14
   1.4 流体性质 / 14
      1.4.1 可压缩性 / 14
      1.4.2 表面张力 / 17
      1.4.3 黏度 / 20

第 2 章
流体静力学 / 27

   2.1 某一点的压力 / 27
   2.2 压力场基本方程 / 28
   2.3 静止流体中的压力变化 / 31
      2.3.1 不可压缩流体 / 31
      2.3.2 可压缩流体 / 34
   2.4 标准大气 / 35
   2.5 浮力和稳定性 / 37
      2.5.1 阿基米德原理 / 37
      2.5.2 稳定性 / 40
   2.6 压力的测量 / 42
   2.7 压力测量法 / 44

2.7.1 Piezometer Tube / 45
2.7.2 U-Tube Manometer / 46
2.7.3 Inclined-Tube Manometer / 50

2.7.1 测压管 / 45
2.7.2 U形管压力计 / 46
2.7.3 倾斜管压力计 / 50

Chapter 3
Fluid Kinematics / 54

3.1 The Velocity Field / 54
   3.1.1 Eulerian Method and Lagrangian Method / 55
   3.1.2 One-, Two- and Three-Dimensional Flows / 56
   3.1.3 Steady and Unsteady Flows / 57
   3.1.4 Streamlines, Streaklines and Pathlines / 58
3.2 The Acceleration Field / 60
   3.2.1 The Material Derivative / 60
   3.2.2 Unsteady Effects / 62
   3.2.3 Convective Effects / 63
3.3 Fluid Element Kinematics / 65
   3.3.1 Linear Motion and Deformation / 66
   3.3.2 Angular Motion and Deformation / 67
3.4 System and Control Volume / 69
3.5 Reynolds Transport Theorem / 71
   3.5.1 Derivation of the Reynolds Transport Theorem / 73
   3.5.2 Relationship Between Reynolds Transport Theorem and Material Derivative / 78

第3章
流体运动学 / 54

3.1 速度场 / 54
   3.1.1 欧拉法和拉格朗日法 / 55
   3.1.2 一维、二维和三维流动 / 56
   3.1.3 定常和非定常流动 / 57
   3.1.4 流线、纹线和迹线 / 58
3.2 加速度场 / 60
   3.2.1 物质导数 / 60
   3.2.2 非定常作用 / 62
   3.2.3 对流作用 / 63
3.3 流体微团运动 / 65
   3.3.1 线运动和变形 / 66
   3.3.2 角运动和变形 / 67
3.4 系统和控制体 / 69
3.5 雷诺输运定理 / 71
   3.5.1 雷诺输运定理的推导 / 73
   3.5.2 雷诺输运定理与物质导数的关系 / 78

Chapter 4
Finite Control Volume Analysis of Fluid Flow / 82

4.1 The Continuity Equation / 82
   4.1.1 Derivation of the Continuity Equation / 82
   4.1.2 Application of the Continuity Equation / 85
4.2 The Momentum Equation / 88
   4.2.1 Derivation of the Momentum Equation / 88
   4.2.2 Application of the Momentum Equation / 89
4.3 Moment-of-Momentum Equation / 91

第4章
流体流动的有限控制体分析 / 82

4.1 连续性方程 / 82
   4.1.1 连续性方程的推导 / 82
   4.1.2 连续性方程的应用 / 85
4.2 动量方程 / 88
   4.2.1 动量方程的推导 / 88
   4.2.2 动量方程的应用 / 89
4.3 动量矩方程 / 91

4.3.1 Derivation of the Moment-of-Momentum Equation / 91
4.3.2 Application of the Moment-of-Momentum Equation / 93
4.4 The Energy Equation / 96
4.4.1 Derivation of the Energy Equation / 96
4.4.2 Application of the Energy Equation / 98
4.4.3 The Bernoulli Equation / 100

Chapter 5
Differential Analysis of Fluid Flow / 108

5.1 Conservation of Mass / 108
5.1.1 Continuity Equation in Differential Form / 109
5.1.2 Continuity Equation in Cylindrical Coordinates / 112
5.2 Conservation of Momentum / 112
5.2.1 Forces Acting on the Differential Element / 113
5.2.2 Equations of Motion / 116
5.3 Viscous Flow / 118
5.3.1 Stress-Deformation Relationships / 118
5.3.2 The Naiver-Stokes Equations / 119
5.4 Solutions for Viscous Incompressible Flow / 120
5.4.1 Steady, Laminar Flow Between Fixed Parallel Plates / 120
5.4.2 Steady, Laminar Flow in Circular Tubes / 123

Chapter 6
Similitude and Dimensional Analysis / 129

6.1 Similitude / 129
6.2 Similarity Laws / 131
6.3 Dimensional Analysis / 134
6.3.1 Dimensional Homogeneity Principle / 134

4.3.1 动量矩方程的推导 / 91
4.3.2 动量矩方程的应用 / 93
4.4 能量方程 / 96
4.4.1 能量方程的推导 / 96
4.4.2 能量方程的应用 / 98
4.4.3 伯努利方程 / 100

第 5 章
流体流动的微分分析 / 108

5.1 质量守恒 / 108
5.1.1 微分形式的连续性方程 / 109
5.1.2 柱坐标系中的连续性方程 / 112
5.2 动量守恒 / 112
5.2.1 作用在微元上的力 / 113
5.2.2 运动方程 / 116
5.3 黏性流动 / 118
5.3.1 应力-变形关系 / 118
5.3.2 纳维-斯托克斯方程 / 119
5.4 黏性不可压缩流动的求解 / 120
5.4.1 固定平板间的定常层流流动 / 120
5.4.2 圆管内的定常层流流动 / 123

第 6 章
相似理论和量纲分析 / 129

6.1 相似理论 / 129
6.2 相似准则 / 131
6.3 量纲分析 / 134
6.3.1 量纲和谐原理 / 134

6.3.2 The Rayleigh Method / 135	6.3.2 瑞利法 / 135
6.3.3 The Buckingham's Π Theorems / 137	6.3.3 白金汉姆Π定理 / 137
6.3.4 Application of the Buckingham's Π Theorems / 139	6.3.4 Π定理的应用 / 139
6.4 Similitude and Modeling / 143	6.4 相似与模化 / 143
6.4.1 Approximate Model of Fluid Mechanics Problem / 143	6.4.1 流体力学问题的近似模型 / 143
6.4.2 Modeling Example / 148	6.4.2 模化实例 / 148

Chapter 7
Pipe Flow / 154

第7章
管内流动 / 154

7.1 General Characteristics of Pipe Flow / 155	7.1 管内流动的一般特性 / 155
7.2 Laminar Flow in Circular Pipe / 158	7.2 圆管中的层流 / 158
7.3 Turbulent Flow in Circular Pipe / 162	7.3 圆管中的湍流 / 162
7.4 Pressure Head Losses in Circular Pipe / 166	7.4 圆管中的压头损失 / 166
7.4.1 Mechanism of Flow Resistance / 166	7.4.1 流动阻力产生的机理 / 166
7.4.2 Classification of Pipe Flow Resistances / 169	7.4.2 管内流动阻力的分类 / 169
7.4.3 Calculation of Major Head Loss / 171	7.4.3 主要损失的计算 / 171
7.4.4 Calculation of Minor Head Loss / 180	7.4.4 次要损失的计算 / 180
7.5 Calculation of Head Loss in Pipeline / 190	7.5 管路损失计算 / 190
7.5.1 Equivalent Hydraulic Diameter / 190	7.5.1 当量水力直径 / 190
7.5.2 Head Loss Calculation of Pipe System / 191	7.5.2 管道系统的损失计算 / 191

Chapter 8
Planar Potential Flow / 202

第8章
平面势流 / 202

8.1 Potential Function and Stream Function / 202	8.1 势函数与流函数 / 202
8.1.1 Potential Function / 202	8.1.1 势函数 / 202
8.1.2 Stream Function / 205	8.1.2 流函数 / 205
8.2 Simple Potential Flow / 208	8.2 简单势流 / 208
8.2.1 Uniform Linear Flow / 208	8.2.1 均匀直线流动 / 208
8.2.2 Flow in Right-angle Region / 209	8.2.2 直角区域内的流动 / 209
8.2.3 Point Source and Point Sink / 211	8.2.3 点源和点汇 / 211
8.2.4 Pure Circulation Flow / 213	8.2.4 纯环流流动 / 213

8.3　Superposition Principle of Potential Flows / 215

Chapter 9
Flow Around Body Immersed / 225

- 9.1　Overview of Boundary Layer / 225
- 9.2　Characteristics of Boundary Layer / 228
  - 9.2.1　Formation of Boundary Layer / 228
  - 9.2.2　The Laminar and Turbulent Boundary Layers / 231
- 9.3　Boundary Layer Equations / 234
  - 9.3.1　The Governing Equations / 234
  - 9.3.2　Boundary Layer Thickness / 237
- 9.4　Flow Around a Cylinder / 240
  - 9.4.1　Ideal Fluid Flow Around a Cylinder / 240
  - 9.4.2　Viscous Fluid Flow Around a Cylinder / 243
- 9.5　Flow Around a Sphere / 250
  - 9.5.1　Ideal Fluid Flow Around a Sphere / 250
  - 9.5.2　Viscous Fluid Flow Around a Sphere / 253

Reference / 256

8.3　势流叠加原理 / 215

第 9 章
绕物流动 / 225

- 9.1　边界层概述 / 225
- 9.2　边界层特性 / 228
  - 9.2.1　边界层的形成 / 228
  - 9.2.2　层流和湍流边界层 / 231
- 9.3　边界层方程 / 234
  - 9.3.1　控制方程 / 234
  - 9.3.2　边界层厚度 / 237
- 9.4　绕圆柱流动 / 240
  - 9.4.1　理想流体绕圆柱流动 / 240
  - 9.4.2　黏性流体绕圆柱流动 / 243
- 9.5　绕球流动 / 250
  - 9.5.1　理想流体的绕球流动 / 250
  - 9.5.2　黏性流体的绕球流动 / 253

参考文献 / 256

Chapter 1　Introduction of Fluid Mechanics
第1章　流体力学简介

Fluid mechanics is the discipline within the field of applied mechanics concerned with the behavior of fluids at rest or in motion. Fluid mechanics has a wide range of applications, including for mechanical engineering, civil engineering, chemical engineering, biology, and so on.

Many interesting questions can be answered by using relatively simple fluid mechanics ideas. A foundation of the fluid mechanics will be introduced in this chapter.

1.1　Brief History of Fluid Mechanics

Before the study of fluid mechanics, the history of this important engineering science should be reviewed briefly.

Before second century A. D., Archimedes (287—212 B. C.), who is a Greek mathematician and inventor, first expressed the principles of hydrostatics and flotation. However, for the next 1000 years, no significant development has been made for fluid mechanics.

About from the fifteenth century to eighteenth century, a rather continuous series of contributions form the basis of fluid mechanics. Leonardo da Vinci (1452—1519) described many different types of flow phenomena. The work of Galileo Galilei (1564—1642) marked the beginning of experimental mechanics. Since then, a large number of famous scientists, such as Newton, Bernoulli, Euler and d'Alembert, etc., made great contributions in theory and mathematics. Experimental aspects of fluid mechanics were also advanced during this period, but the two different methods, theoretical and experimental, developed along separate paths.

流体力学是应用力学领域中的一门学科，研究流体在静止或运动状态下的行为。流体力学在机械工程、土木工程、化学工程和生物学等领域具有广泛的应用。

许多有趣的问题可以通过利用相对简单的流体力学思想来解答。本章简单介绍流体力学基础知识。

1.1　流体力学简史

学习流体力学前，先简要回顾一下这门重要工程科学的发展史。

公元2世纪以前，古希腊数学家和发明家阿基米德首先阐述了流体静力学和漂浮的基本原理。然而，随后的一千多年中，流体力学几乎未得到任何发展。

从15世纪到18世纪，科学家们的一系列贡献形成了流体力学的基础。达·芬奇描绘了许多不同类型的流动现象，伽利略的工作标志着实验力学的开端。此后，出现了一大批著名的科学家，如：牛顿、伯努利、欧拉和达朗贝尔等，在理论和数学方面作出了重大贡献。这一时期，实验流体力学也得到了发展。但是，理论和实验这两种方法各自沿着不同的路径发展。

During the nineteenth century, the differential equations were used to describe fluid motions, and further contributions were made to both theoretical hydrodynamics and experimental hydraulics. Experimental hydraulics gradually became a free-standing science, and many of those experimental results are still used today.

Both the theoretical hydrodynamics and experimental hydraulics were highly developed at the beginning of the twentieth century, and some attempts were made to unify them. In 1904 Ludwig Prandtl (1857—1953) proposed the concept of boundary layer, which laid the foundation for the unification of the theoretical and experimental aspects of fluid mechanics. Prandtl is generally accepted as the founder of modern fluid mechanics.

Also, during the first decade of the twentieth century, powered flight was first successfully demonstrated with the subsequent vastly increased interest in aerodynamics. The design of aircraft required a degree of understanding of fluid flow and an ability to make accurate predictions of the effect of air flow on bodies, so the development of aerodynamics provided a great stimulus for the rapid development in fluid mechanics.

Main contributions of the pioneers in fluid mechanics are summarized in Table 1.1.

19世纪，微分方程被用于描述流体运动，理论水动力学和实验水力学得到了进一步发展。实验水力学逐渐成为一门独立的科学，许多实验结果至今仍在使用。

20世纪初，理论水动力学和实验水力学均得到高度发展，人们试图将它们统一起来。1904年，普朗特提出了边界层的概念，为流体力学理论与实验研究的统一奠定了基础。他被公认为现代流体力学的奠基人。

此外，20世纪前十年，动力飞行的首次演示成功引起了人们对空气动力学的浓厚兴趣。由于飞机的设计要求对流体流动有一定程度的了解，且要求能准确预测空气对飞行物的作用，所以空气动力学的发展大大推动了流体力学的发展。

表1.1总结了前人在流体力学方面的主要贡献。

Table 1.1　Contributions of the pioneers in fluid mechanics

表1.1　前人在流体力学方面的贡献

Name 人名	Contributions 贡献
Archimedes 阿基米德(287—212 B.C.)	Established elementary principles of buoyancy and flotation 建立了浮力和漂浮的基本原理
Leonardo da Vinci 达·芬奇(1452—1519)	Expressed elementary principle of continuity; observed and sketched many basic flow phenomena; designed hydraulic machinery 表达了连续性的基本原理，观察并描述了许多基本的流动现象，设计了水力机械
Galileo Galilei 伽利略(1564—1642)	Indirectly stimulated experimental hydraulics; revised Aristotelian concept of vacuum 间接促进了实验水力学，修正了亚里士多德真空的概念
Evangelista Torricelli 托里拆利(1608—1647)	Invented barometer 发明了气压计

续表

Name 人名	Contributions 贡献
Blaise Pascal 帕斯卡(1623—1662)	Put forward the law that the fluid can transfer pressure, the so-called Pascal's law 提出流体能传递压力的定律,即所谓帕斯卡定律
Isaac Newton 牛顿(1642—1727)	Explored various aspects of fluid resistance, inertia, viscosity and wave 探索了流体阻力、惯性、黏性和波浪的各个方面
Henri de Pitot 皮托(1695—1771)	Invented the pitot tube used for the measurement of flow velocity 发明了测量流速的皮托管
Daniel Bernoulli 伯努利(1700—1782)	Established the basic equation of fluid dynamics, which is called the Bernoulli equation 建立了流体动力学的基本方程,称为伯努利方程
Leonhard Euler 欧拉(1707—1783)	First explained role of pressure in fluid flow; formulated basic equations of motion; introduced concept of cavitation and principle of centrifugal machinery 首先解释了压力在流体流动中的作用,阐述了基本的运动方程,介绍了离心机械的空化概念和原理
Jean le Rond d'Alembert 达朗贝尔(1717—1783)	Originated notion of velocity and acceleration components, and differential expression of continuity 提出了速度和加速度分量的概念和连续性的微分表示
Giovanni Battista Venturi 文丘里(1746—1822)	Performed tests on various forms of mouthpieces, in particular, conical contractions and expansions 对多种喷口形式进行了试验,尤其是圆锥收缩和扩张形式的喷口
Louis Marie Henri Navier 纳维(1785—1836)	Established the basic equations of fluid motion considering "molecular" forces 建立了考虑分子力的流体运动的基本方程
George Gabriel Stokes 斯托克斯(1819—1903)	Established the basic equation of viscous fluid motion, which is called Navier-Stokes equation 建立了黏性流体运动的基本方程,称为纳维-斯托克斯方程
Ernst Mach 马赫(1838—1916)	One of the pioneers in the field of supersonic aerodynamics 超音速空气动力学领域的先驱
Osborne Reynolds 雷诺(1842—1912)	Discovered the similarity law of flow; introduced a dimensionless number, that is the Reynolds number, as the criterion for judging the flow states 发现流动的相似律,引入无量纲数,即雷诺数,作为判别流态的标准
John William Strutt, Lord Rayleigh 瑞利(1842—1919)	Investigated hydrodynamics of bubble collapse, wave motion, jet instability and dynamic similarity 研究了气泡破裂、波动、射流不稳定性和动力相似性等流体力学问题
Vincenc Strouhal 斯特劳哈尔(1850—1922)	Investigated the phenomenon of "singing wires" 研究了丝线发声现象
Edgar Buckingham 白金汉姆(1867—1940)	Stimulated interest in the United States in the use of dimensional analysis 激发了美国在量纲分析使用方面的兴趣
Moritz Weber 韦伯(1871—1951)	Formulated a capillarity similarity parameter 阐述了毛细相似参数
Ludwig Prandtl 普朗特(1875—1953)	Introduced concept of the boundary layer and is generally considered to be the father of present-day fluid mechanics 引入了边界层的概念,被认为是现代流体力学之父
Lewis Ferry Moody 莫迪(1880—1953)	Provided many innovations in the field of hydraulic machinery. Proposed a method of correlating pipe resistance data 在水力机械领域有许多创新,提出了关于管道阻力数据的相关方法

Name 人名	Contributions 贡献
Theodor Von Kármán 冯卡门 (1881—1963)	One of the recognized leaders of twentieth century fluid mechanics. Provided major contributions to our understanding of surface resistance, turbulence and wake phenomena 被认为是 20 世纪流体力学的领袖之一,为人们理解表面阻力、湍流和尾迹现象作出了重要的贡献
Paul Richard Heinrich Blasius 布拉修斯 (1883—1970)	Provided an analytical solution to the boundary layer equations. Also, demonstrated that pipe resistance was related to the Reynolds number 提供了边界层方程的解析解,并证明了管道阻力与雷诺数有关

1.2 Dimensions and Units

1.2.1 Dimensions

In the study of fluid mechanics, we will deal with a variety of fluid characteristics, so it is necessary to develop a system for describing these characteristics both qualitatively and quantitatively. The qualitative aspect refers to identify the nature or type of the characteristics (such as length, time, stress and velocity), whereas the quantitative aspect provides a numerical measure of the characteristics. The quantitative description requires both a number and a standard by which various quantities can be compared. A standard for time might be a second or hour, for mass a kilogram or gram, and for length a meter or millimeter. Such standards are called units.

The qualitative description is given in terms of certain primary quantities, such as length, L, time, T, mass, M, and temperature, Θ. These primary quantities can then be used to provide a qualitative description of any other secondary quantities: for example, area $\doteq L^2$, velocity $\doteq LT^{-1}$, density $\doteq ML^{-3}$, where the symbol indicates the dimensions of the secondary quantity in terms of the primary quantities. Thus, velocity, v, can be described qualitatively as

$$v \doteq LT^{-1}$$

1.2 量纲和单位

1.2.1 量纲

在流体力学的学习过程中,会遇到流体的各种特性,所以有必要建立一个体系以定性和定量地描述这些特性。定性方面是指特性的本质或类型(如:长度、时间、应力和速度),而定量方面提供了特性的数量。定量描述需要具体的数值和度量衡标准,通过它可以比较不同的量。如时间的度量衡标准可以是秒或小时,质量可以是千克或克,长度可以是米或毫米。这样的度量衡标准称作为单位。

根据主量(如:长度 L、时间 T、质量 M 和温度 Θ)可以进行定性描述,任何其他二次量的定性描述可由主量导出,例如:面积 $\doteq L^2$,速度 $\doteq LT^{-1}$,密度 $\doteq ML^{-3}$,这些符号表示根据主量得到的二次量的量纲。因此,速度 v 可定性地描述为

$$v \doteq LT^{-1}$$

that is, the dimensions of velocity equal length divided by time. The primary quantities are also referred to as basic dimensions.

也就是说，速度的量纲等于长度除以时间。主量也被称为基本量纲。

For most fluid mechanics problems, only the three basic dimensions, L, T and M are usually required. Table 1.2 provides a list of dimensions for common physical quantities.

对于大多数流体力学问题而言，物理量通常只需要 L、T 和 M 三个基本量纲。表 1.2 给出了常用物理量的量纲。

Table 1.2 Dimensions associated with common physical quantities
表 1.2 常用物理量的量纲

Physical quantities 物理量	Dimension 量纲	SI Units 国际单位
Acceleration 加速度	LT^{-2}	m/s^2
Angle 角度	$M^0 L^0 T^0$	1
Angular acceleration 角加速度	T^{-2}	s^{-2}
Angular velocity 角速度	T^{-1}	s^{-1}
Area 面积	L^2	m^2
Density 密度	ML^{-3}	kg/m^3
Energy 能量	$ML^2 T^{-2}$	J
Force 力	MLT^{-2}	N
Length 长度	L	m
Mass 质量	M	kg
Modulus of elasticity 弹性模量	$ML^{-1}T^{-2}$	Pa
Moment of a force 力矩	$ML^2 T^{-2}$	N·m
Momentum 动量	MLT^{-1}	kg·m/s
Power 功率	$ML^2 T^{-3}$	W
Pressure 压强	$ML^{-1}T^{-2}$	Pa
Specific weight 重度	$ML^{-2}T^{-2}$	N/m^3
Strain 应变	$M^0 L^0 T^0$	1
Stress 应力	$ML^{-1}T^{-2}$	MPa
Surface tension 表面张力	MT^{-2}	N/m
Temperature 温度	Θ	K
Time 时间	T	s
Torque 扭矩	$ML^2 T^{-2}$	N·m
Velocity 速度	LT^{-1}	m/s
Viscosity(dynamic) 动力黏度	$ML^{-1}T^{-1}$	$N·s/m^2$
Viscosity(kinematic) 运动黏度	$L^2 T^{-1}$	m^2/s
Volume 容积	L^3	m^3
Work 功	$ML^2 T^{-2}$	J

All theoretically derived equations are dimensionally homogeneous; that is, the dimensions of the left side of the equation must be consistent with those on the right side, and all additive separate terms must have the same dimensions. For example, the equation for the velocity of uniformly accelerated body is

$$v = v_0 + at \qquad (1.1)$$

where v_0 is the initial velocity, a is the acceleration, and t is the time interval. In terms of dimensions the equation is

$$LT^{-1} \doteq LT^{-1} + LT^{-1}$$

and thus Eq. (1.1) is dimensionally homogeneous.

Some equations contain constants having dimensions. For example, for a freely falling body, the traveled distance, d, can be written as

$$d = 4.9 t^2 \qquad (1.2)$$

and the constant must have the dimensions of LT^{-2} if the equation is dimensionally homogeneous. Actually, Eq. (1.2) is a special form of the well-known equation for freely falling bodies.

$$d = \frac{gt^2}{2} \qquad (1.3)$$

where g is the acceleration of gravity. Equation (1.3) is dimensionally homogeneous and valid in any system of units. For $g = 9.8 \text{m/s}^2$, the equation reduces to Eq. (1.2) and thus Eq. (1.2) is valid only for the system of units using meter and second. The concept of dimensions forms the basis for the dimensional analysis, which is discussed in detail in Chapter 6.

【EXAMPLE 1.1】 A pipe locates in the side of a tank as shown in Fig. 1.1. The volume flowrate, Q, of liquid through the pipe can be expressed as

$$Q = 0.61 A \sqrt{2gh}$$

所有理论推导得到的方程都是量纲和谐的,也就是说,方程左边的量纲必须与方程右边的量纲一致,所有加和项具有相同的量纲。例如:匀加速物体的速度方程为

式中,v_0 为初速度;a 为加速度;t 为时间间隔。上式采用量纲表示为

因此,式(1.1)是量纲和谐的。

有些方程中的常数是有量纲的。例如:对于自由落体,下降的距离可以表示为

如果方程是量纲和谐的,那么上式中常数的量纲必须为 LT^{-2}。实际上,式(1.2)是自由落体运动方程的特殊形式。

式中,g 为重力加速度。式(1.3)对任何单位系统都是量纲和谐的。如果 $g = 9.8 \text{m/s}^2$,式(1.3)就变为式(1.2),因此,式(1.2)只对使用米和秒作为单位的系统成立。量纲的概念是量纲分析的基础,将在第 6 章详细讨论。

【例题 1.1】 一圆管位于容器侧面,如图 1.1 所示。液体通过圆管的体积流量可表达为

where A is the area of the pipe, g is the acceleration of gravity, and h is the height of the liquid above the pipe center. Investigate the dimensional homogeneity of this formula.

式中，A 为圆管的面积；g 为重力加速度；h 为圆管中心上方液体的高度。试分析该公式的量纲和谐性。

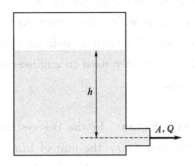

Fig. 1.1 Figure for the example 1.1
图 1.1 例题 1.1 图

SOLUTION The dimensions of the various terms in the equation are $Q \doteq L^3 T^{-1}$, $A \doteq L^2$, $g \doteq LT^{-2}$, $h \doteq L$. These terms, when substituted into the equation, yield the dimensional form

解 公式中 Q 的量纲为 $L^3 T^{-1}$，A 的量纲为 L^2，g 的量纲为 LT^{-2}，h 的量纲为 L。代入上式得

$$(L^3 T^{-1}) \doteq (0.61)(L^2)(2)^{1/2}(LT^{-2})^{1/2}(L)^{1/2}$$

$$(L^3 T^{-1}) \doteq [(0.61)(2)^{1/2}](L^3 T^{-1})$$

it can be seen that both sides of the equation have the same dimensions of $L^3 T^{-1}$, the equation is dimensionally homogeneous, and the numbers [0.61 and $(2)^{1/2}$] are dimensionless. If g is replaced with the value of 9.8 m/s², the equation can be rewritten as

可见，上式两边具有相同的量纲 $L^3 T^{-1}$，该式量纲和谐，0.61 和 $(2)^{1/2}$ 为无量纲的数。如果用 9.8 m/s² 代替 g，公式可重写为

$$Q = 2.7 A (h^{1/2}) \qquad (1.4)$$

A check of the dimensions reveals that

量纲分析表明

$$L^3 T^{-1} \doteq (2.7)(L^{5/2})$$

and, therefore, the equation expressed as Eq. (1.4) can only be dimensionally correct if the number 2.7 has the dimensions of $L^{1/2} T^{-1}$. Whenever a number appearing in an equation or formula has dimensions, it means that the specific value of the number will depend on the system of units used.

因此，只有当 2.7 的量纲为 $L^{1/2} T^{-1}$ 时，式(1.4)的量纲才正确。当一个方程或公式中的数具有量纲时，这就意味着，这个数的具体值取决于所用的单位系统。

1.2.2 Units

In addition to the qualitative description of the various quantities, it is generally necessary to have a quantitative measure of any given quantity. For example, if the width of a desk is 2 units wide, the statement has no meaning until the unit of length is defined. Now three systems of units commonly used in engineering are introduced as follows.

British Gravitational (BG) System In the BG system, the unit of length is the foot (ft), the unit of time is the second (s), the unit of force is the pound (lb), and the unit of temperature is the degree Fahrenheit (°F) or the unit of thermodynamic temperature is the degree Rankine (°R). The unit of mass, called the slug, is defined from Newton's second law (force = mass×acceleration) as 1lb=(1slug)(1ft/s^2). This relationship indicates that a force of 1lb acting on a mass of 1slug will give the mass an acceleration of 1ft/s^2.

English Engineering (EE) System In the EE system, units for force and mass are defined independently; thus special care must be exercised when using this system in conjunction with Newton's second law. The basic unit of mass is the pound mass (lbm), the unit of force is the pound (lb). It is also common practice to use the notation, lbf, to indicate pound force. The unit of length is the foot (ft), the unit of time is the second (s), and the thermodynamic temperature scale is the degree Rankine (°R).

International System (SI) In 1960 the Eleventh General Conference on Weights and Measures formally adopted the International System of Units (termed SI) as the international standard. This system has been widely adopted worldwide. In SI the unit of length is the meter (m), the unit of time is the second (s), the unit of mass is the kilogram (kg),

1.2.2 单位

除了定性地描述以外，对任意给定的量，通常还要定量测量。例如：一张课桌的宽度为2个单位宽，这样的描述是没有意义的，除非定义了长度的单位。工程上常用的三种单位系统介绍如下。

BG单位系统 在BG单位系统中，长度的单位是英尺（ft），时间的单位是秒（s），力的单位是磅（lb），温度的单位是华氏度（°F）或者热力学温度的单位兰金度（°R），两者的关系为°R=°F+459.67。质量的单位称为斯勒格，根据牛顿第二定律定义为1lb=(1slug)(1ft/s^2)。该式表明，1lb力作用在1slug的质量上，产生1ft/s^{2}的加速度。

EE单位系统 在EE单位系统中，力和质量分别定义。因此，在结合牛顿第二定律使用这个单位系统时应特别注意。质量的基本单位是质量磅（lbm），力的单位是磅（lb），通常也使用lbf这个符号来表示力磅。长度的单位是英尺（ft），时间的单位是秒（s），热力学温度单位是兰金度（°R）。

SI单位系统 1960年，第十一届度量衡会议正式采用国际单位系统（SI）作为国际标准，该系统被各国广泛采用。在SI系统中，长度的单位是米（m），时间的单位是秒（s），质量的单位是千克（kg），温度的单位是开尔文（K）。开尔文温

and the unit of temperature is the Kelvin (K). The Kelvin temperature scale is an absolute scale. The relationship between K and ℃ is K=℃+273.15.

In this book the SI unit system will primarily be used for units. The use of incorrect units will lead to huge errors in problem solutions. Get in the habit of using a consistent system of units throughout a given solution.

【EXAMPLE 1.2】 A box having a total mass of 36kg rests on the floor of an elevator as shown in Fig. 1.2. Determine the force (in Newton) that the box exerts on the floor when the elevator is accelerating upward at $6ft/s^2$.

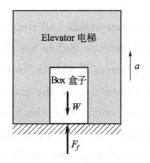

Fig. 1.2　Figure for the example 1.2

SOLUTION　A free-body diagram of the box is shown in Fig. 1.2 where W is the weight of the box, and F_f is the reactive force of the floor on the box. Application of Newton's second law of motion to this body gives

$$\sum \mathbf{F} = m\mathbf{a}$$

or

$$F_f - W = ma \quad (1.5)$$

Since $W=mg$, Eq. (1.5) can be written as

$$F_f = m(g+a) \quad (1.6)$$

Before substituting any number into Eq. (1.6), we must decide on a system of units, and then be sure all of the data are expressed in these units. Since we want

F_f in Newton, we will use SI units so that

$$F_f = 36\text{kg}[9.81\text{m/s}^2 + (6\text{ft/s}^2)(0.3048\text{m/ft})] = 419\text{kg} \cdot \text{m/s}^2$$

Since $1\text{N} = 1\text{kg} \cdot \text{m/s}^2$, it follows that

$$F_f = 419\text{N}$$

The force shown on the free-body diagram is the force of the floor on the box so that the force the box exerts on the floor is equal in magnitude but opposite in direction.

1.3 Definition of a Fluid

1.3.1 Continuity Hypothesis

What is the difference between solids and fluids? We have a vague idea of the difference. Solids are "hard" and not easily deformed, whereas fluids are "soft" and easily deformed. From the perspective of the molecular structure of materials, we commonly think solids have densely space molecules with large intermolecular cohesive forces that allow the solid to maintain its shape. However, for liquids, the molecules are spaced farther apart, the intermolecular forces are smaller than the forces for solids, and the molecules have more freedom of movement. Thus, liquids can be easily deformed (but not easily compressed). Gases have even greater molecular spacing and freedom of motion. The intermolecular forces are almost negligible, and therefore gases are easily deformed and compressed.

Although the differences between solids and fluids can be explained qualitatively on the basis of molecular structure, according to the deformation characteristics under the action of an external load, a clearer distinction between them can be made. Specifically, a fluid is defined as a substance that deforms continuously when acted on by a shearing stress of any magnitude. Once the tangential force acts on the surface, it produces shear stress. When common solids such as steel or other metals are acted on by a shearing stress, they will produce a very small deformation, but they will not continuously deform (flow). However, common fluids such as water, oil and air satisfy the definition of a fluid—that is, they will flow when acted on by a shearing stress. Some materials, such as slurries, tar, toothpaste, and so on, are not easily classified since they will behave as a solid if the applied shearing stress is small, but if the stress exceeds some critical values, the substance will flow. All the fluids in this book will conform to the definition of a fluid given previously.

Although the molecular structure of fluids is important in distinguishing one fluid from another, it is not possible to study the behavior of individual molecules when trying to describe the behavior of fluids. Inversely, the behavior of a fluid is characterized by a statistical average of a small volume containing a large number of molecules. The volume is small compared with the physical dimensions of the system of interest, but large compared with the distance between molecules.

Since the distance between molecules is typically very small, this is a reasonable way to describe the behavior of a fluid. For gases at normal pressures and temperatures, the distance is on the order of 10^{-6} mm and for liquids it is on the order of 10^{-7} mm. The number of molecules per cubic millimeter is on the order of 10^{18} for gases and 10^{21} for liquids. Obviously, even in a very small volume, the number of molecules is also very large, so the statistical average is undoubtedly reasonable. Thus, it is assumed that all the fluid characteristics (pressure, velocity, etc.) vary continuously throughout the fluid, that is, the fluid is treated as a continuum. The continuum hypothesis is no longer acceptable in the study of rarefied gases since in this case the distance between gas molecules can become very large.

体（如：钢铁或其他金属）受到切应力的时候，会产生较小的变形，但不会连续变形而流动起来。然而，常见的流体（如：水、油或者空气）满足流体的定义，也就是说只要受到切应力的作用它们就会流动起来。有些材料（如：泥浆、沥青和牙膏等）很难进行分类，因为当切应力较小时它们表现为固体，当切应力超过某一临界值又会流动起来。本书中的流体均符合前面流体的定义。

尽管流体的分子结构在区分一种流体和另一种流体时是重要的，但在描述流体行为时，不可能研究单个分子的行为。相反，流体的行为用包含大量分子的一小块体积内的统计平均值来表征。这个体积相对于所研究系统的物理尺度属于小量，但相对于分子间距是足够大的。

因为分子间的距离通常非常小，这是一种描述流体的行为的合理方法。常温常压下，气体分子间距离的数量级为 10^{-6} mm，液体的数量级为 10^{-7} mm。每立方毫米气体分子的数量的数量级为 10^{18} 数量级，液体分子的数量级为 10^{21}。显然，即便在很小的体积内，分子数量也非常大，故采用统计平均值无疑是合理的。因此，假设所有流体的特性（如：压力、速度等）都是连续变化的，即，流体被当作连续介质来处理。对于稀薄气体，气体分子间的距离很大，连续性假设不成立。

1.3.2 Density

The density of a fluid, designated by the Greek symbol ρ, is defined as its mass per unit volume. In the SI unit system its unit is kg/m^3.

The value of density can vary in a wide range for different fluids, but for liquids, variations in pressure and temperature generally have only a small effect on the value of ρ. Unlike liquids, the density of a gas is strongly influenced by both pressure and temperature. The change in the density of water with variations in temperature is illustrated in Fig. 1.3. Table 1.3 lists values of density for several common liquids.

The specific volume, v, is the volume per unit mass and is therefore the reciprocal of the density—that is

1.3.2 密度

流体密度定义为单位体积内流体的质量,用希腊符号 ρ 表示,在 SI 单位系统中其单位是 kg/m^3。

对于不同的流体,密度值在很宽的范围内变化,但是对于液体,压力和温度对密度 ρ 的影响较小。和液体不同,气体的密度受压力和温度的影响较大。水的密度随温度的变化如图 1.3 所示。表 1.3 列出了几种常见液体的密度值。

比容 v 表示单位质量所具有的体积,它是密度的倒数,即

$$v = \frac{1}{\rho} \qquad (1.7)$$

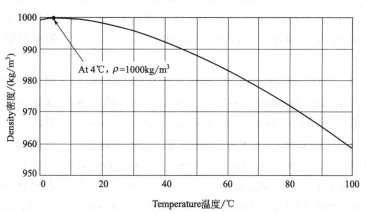

Fig. 1.3 Variation of water density with temperature

图 1.3 水的密度随温度的变化

Table 1.3 Physical properties of some common liquids

表 1.3 常见液体的物理性质

Liquid 液体	Temperature 温度 $T/℃$	Density 密度 $\rho/(kg/m^3)$	Specific Weight 重度 $\gamma/(kN/m^3)$	Dynamic Viscosity 动力黏度 $\mu/(N \cdot s/m^2)$	Kinematic Viscosity 运动黏度 $\nu/(m^2/s)$	Surface Tension 表面张力 $\sigma/(N/m)$	Bulk Modulus 体积模量 $E_V/(N/m^2)$
Carbon tetrachloride 四氯化碳	20	1590	15.6	9.58×10^{-4}	6.03×10^{-7}	2.69×10^{-2}	1.31×10^9

续表

Liquid 液体	Temperature 温度 $T/℃$	Density 密度 $\rho/(kg/m^3)$	Specific Weight 重度 $\gamma/(kN/m^3)$	Dynamic Viscosity 动力黏度 $\mu/(N\cdot s/m^2)$	Kinematic Viscosity 运动黏度 $\nu/(m^2/s)$	Surface Tension 表面张力 $\sigma/(N/m)$	Bulk Modulus 体积模量 $E_V/(N/m^2)$
Ethyl alcohol 乙醇	20	789	7.74	1.19×10^{-3}	1.51×10^{-6}	2.28×10^{-2}	1.06×10^9
Gasoline 汽油	15.6	680	6.67	3.1×10^{-4}	4.6×10^{-7}	2.2×10^{-2}	1.30×10^9
Glycerin 甘油	20	1260	12.4	1.50	1.19×10^{-3}	6.33×10^{-2}	4.52×10^9
Mercury 水银	20	13600	133	1.57×10^{-3}	1.15×10^{-7}	4.66×10^{-1}	2.85×10^{10}
Seawater 海水	15.6	1030	10.1	1.20×10^{-3}	1.17×10^{-6}	7.34×10^{-2}	2.34×10^9
Water 淡水	15.6	999	9.80	1.12×10^{-3}	1.12×10^{-6}	7.34×10^{-2}	2.15×10^9

1.3.3 Specific Weight

The specific weight of a fluid, designated by the Greek symbol γ, is defined as its weight per unit volume. Thus, the relationship between the specific weight and the density is

$$\gamma = \rho g \qquad (1.8)$$

where g is the acceleration of gravity. In the SI unit system, the unit of γ is N/m^3. Values of specific weight for several common liquids are listed in Table 1.3. More complete physical properties for water can be found in Table 1.4.

1.3.3 重度

重度定义为单位体积所具有的重量，用希腊符号 γ 表示。因此，重度与密度之间的关系为

式中，g 为重力加速度。在 SI 单位系统中，γ 的单位是 N/m^3。常见液体的重度如表 1.3 所列，水的详细物理性质如表 1.4 所列。

Table 1.4 Physical properties of water
表 1.4 水的物理性质

Temperature 温度 $T/℃$	Density 密度 $\rho/(kg/m^3)$	Specific Weigh 重度 $\gamma/(kN/m^3)$	Dynamic Viscosity 动力黏度 $\mu/(N\cdot s/m^2)$	Kinematic Viscosity 运动黏度 $\nu/(m^2/s)$	Surface Tension 表面张力 $\sigma/(N/m)$
0	999.9	9.806	1.787×10^{-3}	1.787×10^{-6}	7.56×10^{-2}
5	1000.0	9.807	1.519×10^{-3}	1.519×10^{-6}	7.49×10^{-2}
10	999.7	9.804	1.307×10^{-3}	1.307×10^{-6}	7.42×10^{-2}
20	998.2	9.789	1.002×10^{-3}	1.004×10^{-6}	7.28×10^{-2}
30	995.7	9.765	7.975×10^{-4}	8.009×10^{-7}	7.12×10^{-2}
40	992.2	9.731	6.529×10^{-4}	6.580×10^{-7}	6.96×10^{-2}
50	988.1	9.690	5.468×10^{-4}	5.534×10^{-7}	6.79×10^{-2}
60	983.2	9.642	4.665×10^{-4}	4.745×10^{-7}	6.62×10^{-2}
70	977.8	9.589	4.042×10^{-4}	4.134×10^{-7}	6.44×10^{-2}
80	971.8	9.530	3.547×10^{-4}	3.650×10^{-7}	6.26×10^{-2}
90	965.3	9.467	3.147×10^{-4}	3.260×10^{-7}	6.08×10^{-2}
100	958.4	9.399	2.818×10^{-4}	2.940×10^{-7}	5.89×10^{-2}

1.3.4 Specific Gravity

The specific gravity of a fluid, designated as SG, is defined as the ratio of the density of the fluid to the density of water at 4℃. It can be expressed as

$$SG = \frac{\rho}{\rho_{H_2O,4℃}} \quad (1.9)$$

and since it is the ratio of densities, the value of SG does not depend on the system of units used.

It is clear that density, specific weight and specific gravity are all interrelated, and from any one of the three the others can be calculated.

1.4 Fluid Properties

1.4.1 Compressibility

(1) Bulk Modulus

The bulk modulus, designated as E_V, is commonly used to characterize compressibility, which is defined as

$$E_V = -\frac{dp}{dV/V} \quad (1.10)$$

where dp is the differential change in pressure needed to create a differential change in volume, dV, of a volume V. The negative sign represents that an increase in pressure will cause a decrease in volume. Since a decrease in volume of a given mass will result in an increase in density, Eq. (1.10) can also be expressed as

$$E_V = \frac{dp}{d\rho/\rho} \quad (1.11)$$

The bulk modulus (also is referred to as the bulk modulus of elasticity) has dimensions of pressure, FL^{-2}. In SI unit system its unit is N/m^2 (Pa) and values of E_V for common liquids are listed in Table 1.3. Large

1.3.4 比重

流体的比重 SG 定义为流体密度与 4℃下水的密度之比，可表示为

由于比重是密度的比值，所以它与使用何种单位系统无关。

很明显，密度、重度和比重三者相互关联，已知其中一个，即可计算出其他两个。

1.4 流体性质

1.4.1 可压缩性

(1) 体积模量

体积模量 E_V 用来表征流体的可压缩性，定义为

式中，dp 表示引起体积 V 产生微小变化 dV 所需要的压力变化；负号表示压力的增加将引起体积的减小。因为对于给定的质量，体积的减小将导致密度的增加，所以式(1.10) 也可以写成

体积模量（也称作弹性体积模量）具有压力的量纲，FL^{-2}。在 SI 单位系统中，它的单位为 N/m^2 (Pa)。常见液体的体积模量见表1.3，体

values for the bulk modulus indicate that the fluid is relatively incompressible. Since a large pressure change can only cause a small change in volume, liquids can be considered as incompressible fluids for most practical engineering applications.

(2) Compression and Expansion of Gases

Gases are much easier to compress than liquids. The changes in gas density are directly related to changes in pressure and temperature. Their relationship can be expressed as

$$p = \rho R T \quad (1.12)$$

where p is the absolute pressure, ρ is the density, T is the thermodynamic temperature, and R is the gas constant. Equation (1.12) is commonly termed the equation of state for an ideal gas.

When gases are compressed or expanded the relationship between pressure and density depends on the nature of the process. If the compression or expansion takes place under constant temperature conditions (isothermal process), then

$$\frac{p}{\rho} = \text{constant 常数} \quad (1.13)$$

If the compression or expansion is frictionless and no heat is exchanged with the surroundings (isentropic process), then

$$\frac{p}{\rho^k} = \text{constant 常数} \quad (1.14)$$

where k is the ratio of the specific heat capacity at constant pressure, c_p, to the specific heat capacity at constant volume, c_V. The relationship between the two specific heats capacity and the gas constant, R, is $R = c_p - c_V$. The pressure in both Eq. (1.13) and Eq. (1.14) must be absolute pressure. Values of k for some common gases are listed in Table 1.5, and for air in a range of temperatures, in Table 1.6.

积模量大说明流体不易压缩。由于很大的压力变化只能引起较小的体积改变，通常在实际工程应用中认为液体是不可压缩流体。

(2) 气体的压缩与膨胀

与液体相比，气体容易压缩得多。气体密度的变化与压力和温度的变化直接相关。它们之间的关系可表示为

式中，p 为绝对压力；ρ 为密度；T 为热力学温度；R 为气体常数。方程 (1.12) 通常称为理想气体状态方程。

气体压缩或膨胀时，压力和密度之间的关系取决于变化过程的性质。若压缩或膨胀在等温条件下发生（等温过程），则

若压缩或膨胀过程无摩擦且与环境间无热交换（等熵过程），则

式中，k 为定压比热容 c_p 与定容比热容 c_V 之比，这两个比热容与气体常数 R 的关系为 $R = c_p - c_V$。式(1.13) 和式(1.14) 中的压力必须为绝对压力。常见气体的 k 值如表 1.5 所列，一定温度范围内空气的 k 值如表 1.6 所列。

Table 1.5　Physical properties of some common gases at standard atmospheric pressure
表1.5　标准大气压下常见气体的物理性质

Gas　气体	Temperature 温度 $T/°C$	Density 密度 $\rho/(kg/m^3)$	Specific Weight 重度 $\gamma/(N/m^3)$	Dynamic Viscosity 动力黏度 $\mu/(N \cdot s/m^2)$	Kinematic Viscosity 运动黏度 $\nu/(m^2/s)$	Gas Constant 气体常数 $R/[J/(kg \cdot K)]$	Specific Heat Ratio 比热比 $k/(—)$
Air(standard)　空气	15	1.23	1.20×10^1	1.79×10^{-5}	1.46×10^{-5}	2.869×10^2	1.40
Carbon dioxide　二氧化碳	20	1.83	1.80×10^1	1.47×10^{-5}	8.03×10^{-6}	1.889×10^2	1.30
Helium　氦气	20	1.66×10^{-1}	1.63	1.94×10^{-5}	1.15×10^{-4}	2.077×10^3	1.66
Hydrogen　氢气	20	8.38×10^{-2}	8.22×10^{-1}	8.84×10^{-6}	1.05×10^{-4}	4.124×10^3	1.41
Methane　甲烷	20	6.67×10^{-1}	6.54	1.10×10^{-5}	1.65×10^{-5}	5.183×10^2	1.31
Nitrogen　氮气	20	1.16	1.14×10^1	1.76×10^{-5}	1.52×10^{-5}	2.968×10^2	1.40
Oxygen　氧气	20	1.33	1.30×10^1	2.04×10^{-5}	1.53×10^{-5}	2.598×10^2	1.40

Table 1.6　Physical properties of air at standard atmospheric pressure
表1.6　标准大气压下空气的物理性质

Temperature 温度 $T/°C$	Density 密度 $\rho/(kg/m^3)$	Specific Weigh 重度 $\gamma/(N/m^3)$	Dynamic Viscosity 动力黏度 $\mu/(N \cdot s/m^2)$	Kinematic Viscosity 运动黏度 $\nu/(m^2/s)$	Specific Heat Ratio 比热比 $k/(—)$
−40	1.514	14.85	1.57×10^{-5}	1.04×10^{-5}	1.401
−20	1.395	13.68	1.63×10^{-5}	1.17×10^{-5}	1.401
0	1.292	12.67	1.71×10^{-5}	1.32×10^{-5}	1.401
5	1.269	12.45	1.73×10^{-5}	1.36×10^{-5}	1.401
10	1.247	12.23	1.76×10^{-5}	1.41×10^{-5}	1.401
15	1.225	12.01	1.80×10^{-5}	1.47×10^{-5}	1.401
20	1.204	11.81	1.82×10^{-5}	1.51×10^{-5}	1.401
25	1.184	11.61	1.85×10^{-5}	1.56×10^{-5}	1.401
30	1.165	11.43	1.86×10^{-5}	1.60×10^{-5}	1.400
40	1.127	11.05	1.87×10^{-5}	1.66×10^{-5}	1.400
50	1.109	10.88	1.95×10^{-5}	1.76×10^{-5}	1.400
60	1.060	10.40	1.97×10^{-5}	1.86×10^{-5}	1.399
70	1.029	10.09	2.03×10^{-5}	1.97×10^{-5}	1.399
80	0.9996	9.803	2.07×10^{-5}	2.07×10^{-5}	1.399
90	0.9721	9.533	2.14×10^{-5}	2.20×10^{-5}	1.398
100	0.9461	9.278	2.17×10^{-5}	2.29×10^{-5}	1.397
200	0.7461	7.317	2.53×10^{-5}	3.39×10^{-5}	1.390
300	0.6159	6.040	2.98×10^{-5}	4.84×10^{-5}	1.379
400	0.5243	5.142	3.32×10^{-5}	6.34×10^{-5}	1.368
500	0.4565	4.477	3.64×10^{-5}	7.97×10^{-5}	1.357
1000	0.2772	2.719	5.04×10^{-5}	1.82×10^{-4}	1.321

The bulk modulus for gases can be determined by obtaining the derivative from Eq. (1.13) or Eq. (1.14) and substituting the results into Eq. (1.11). It follows that for an isothermal process

$$E_V = p \quad (1.15)$$

and for an isentropic process

$$E_V = kp \quad (1.16)$$

The bulk modulus in both cases varies directly with pressure. The effects of compressibility on fluid behavior should be considered in the study of gases. However, gases can often be treated as incompressible fluids if the changes in pressure are very small.

1.4.2 Surface Tension

At the interface between a liquid and a gas, or between two immiscible liquids, forces develop on the liquid surface. The forces cause the surface to behave as if it were a "membrane" stretched over the liquid. Although such a membrane is not actually present, this conceptual analogy can help to explain several commonly observed phenomena. For example, an aluminum coin will float on water; small droplets of mercury will form into spheres.

These surface phenomena are due to the unbalanced cohesive forces acting on the liquid molecules at the fluid surface. The attractions of molecules in the interior of the fluid are equal in all directions. However, molecules along the surface are subjected to a net force toward the interior. The intensity of the molecular attraction per unit length along any line in the surface is called the surface tension and is designated by the Greek symbol σ. For a given liquid the surface tension depends on temperature as well as the other fluid, which it is in contact with at the interface. The dimensions of surface tension are FL^{-1} and the unit is N/m. Values of surface tension for some common liquids in contact with air are given in Table 1.3 and Table 1.4.

对式(1.13)和式(1.14)分别求导数,并代入式(1.11)得到气体的体积模量。对于等温过程

对于等熵过程

这两种情况下,体积模量均随压力的变化而变化。研究气体时应该考虑其可压缩性对流体行为的影响,然而,当压力变化很小时,通常可当作不可压缩流体来处理。

1.4.2 表面张力

在气-液或者不相溶的液-液交界面上,形成了作用在液体表面上的力,它使得液体表面好像铺有一层膜。尽管这样的膜并不存在,但这样的概念类比可以帮助解释一些常见的现象,例如:铝质硬币可以浮在水面上,水银滴会聚集成球。

这些表面现象是由于流体表面上的液体分子不平衡的内聚力引起的。液体内部分子所受的吸引力在各个方向上相等,而表面的分子只受到指向液体内部的作用力。沿表面任何一条线的单位长度的分子吸引力强度称为表面张力,用希腊符号σ表示。对于给定的液体,表面张力与温度以及与之接触的流体的有关。表面张力的量纲为FL^{-1},单位为N/m。常见液体与空气接触的表面张力见表1.3和表1.4。

The pressure inside a drop of fluid can be calculated using the free-body diagram in Fig. 1.4. If the spherical drop is cut in half, the force developed around the edge due to surface tension is $2\pi R\sigma$. This force must be balanced by the pressure difference, Δp, between the internal pressure, p_i, and the external pressure, p_e, acting over the circular area, πR^2. Thus

液滴内的压力可以用图 1.4 所示的受力图进行计算，球形液滴切成两半，由于表面张力作用在液滴边缘的力为 $2\pi R\sigma$，它与内外压差 Δp（内压 p_i，外压 p_e）作用在圆形面积 πR^2 上形成的力平衡。因此

$$2\pi R\sigma = \Delta p \pi R^2 \quad \text{or} \quad \Delta p = p_i - p_e = \frac{2\sigma}{R} \qquad (1.17)$$

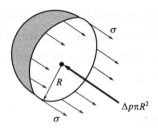

Fig. 1.4　Forces acting on one-half of a liquid drop
图 1.4　作用在半个液滴上的力

It is apparent from this result that the pressure inside the drop is larger than the ambient pressure.

可见，液滴内部的压力大于其环境压力。

The level of a liquid in a capillary tube is related to surface tension. If a capillary tube is inserted into water, the water level in the tube is higher than the water level outside the tube as illustrated in Fig. 1.5(a). For this case, the attraction (adhesion) between the wall of the tube and liquid molecules is strong enough to overcome the mutual attraction (cohesion) of the molecules and pulls them up the wall. Therefore, the liquid is called to wet the solid surface.

毛细管内液面的高低与表面张力有关。将毛细管插入水中，管内液面高于管外液面，如图 1.5(a) 所示。对于这种情况，管壁与液体分子间的吸附力足以克服液体分子间的内聚力，从而把分子沿管壁上拉，因此称液体润湿固体表面。

The height, h, is governed by the value of the surface tension, σ, the tube radius, R, the specific weight of the liquid, γ, and the angle of contact, θ, between the liquid and tube. From the free-body diagram shown in Fig. 1.5(b), it can be seen that, in the vertical direction, the force due to the surface tension is equal to $2\pi R\sigma \cos\theta$ and the weight is $\gamma \pi R^2 h$ and these two forces must maintain balance. Thus

上升的高度 h 由表面张力 σ、管径 R、液体重度 γ 以及液体和管子的接触角 θ 决定。由图 1.5(b) 所示的受力图可见，在竖直方向上，由表面张力引起的力等于 $2\pi R\sigma \cos\theta$，重力等于 $\gamma \pi R^2 h$，二者必须保持平衡，因此

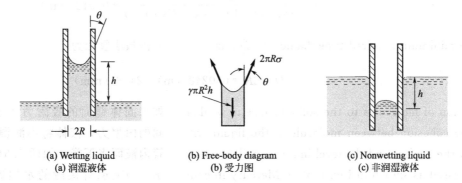

(a) Wetting liquid　(a) 润湿液体
(b) Free-body diagram　(b) 受力图
(c) Nonwetting liquid　(c) 非润湿液体

Fig. 1.5　Effect of capillary action in small tubes
图 1.5　细管内的毛细作用

$$\gamma \pi R^2 h = 2\pi R \sigma \cos\theta$$

| so that the height is given by the Eq. (1.18) | 故高度可由式(1.18) 得到 |

$$h = \frac{2\sigma\cos\theta}{\gamma R} \quad (1.18)$$

The angle of contact is related to both the liquid and the surface. It can be seen from Eq. (1.18) that the height is inversely proportional to the tube radius, and therefore the rise of a liquid in a tube due to the capillary action becomes increasingly pronounced as the tube radius is decreased.

接触角与液体的种类和接触表面有关。由式(1.18) 可见，液面的上升高度与毛细管管径成反比，因此随着管径的减小，毛细作用引起的液体上升越明显。

【EXAMPLE 1.3】 If the rise of water at 20℃ in a tube due to capillary action is less than 1.2mm, what diameter of the glass tube is required?

【例题 1.3】 如果 20℃ 的水在玻璃管内由于毛细作用的升高量小于 1.2mm，玻璃管的管径为多大？

SOLUTOIN From Eq. (1.18)

解　由式(1.18)，有

$$h = \frac{2\sigma\cos\theta}{\gamma R}$$

so that

因此

$$R = \frac{2\sigma\cos\theta}{\gamma h}$$

For water at 20℃ (from Table 1.4), $\sigma = 0.0728$N/m and $\gamma = 9.789$kN/m^3. Since $\theta \approx 0°$ it follows that for $h = 1.2$mm

对于 20℃ 的水，由表 1.4 可得，$\sigma = 0.0728$N/m，$\gamma = 9.789$kN/m^3。由于 $\theta \approx 0°$，当 $h = 1.2$mm 时，有

$$R = \frac{2 \times 0.0728 \times \cos 0°}{(9.789 \times 10^3) \times (1.2 \times 10^{-3})} = 0.0124 \text{ (m)}$$

and the minimum required tube diameter, D, is 所需最小管径为

$$D = 2R = 0.0248 \text{ (m)} = 24.8 \text{ (mm)}$$

If adhesion of molecules to the solid surface is weaker than the cohesion between molecules, the liquid will not wet the surface and the level in a tube will actually be depressed as shown in Fig. 1.5(c). Mercury in contact with a glass tube is a typical example of a nonwetting liquid. For nonwetting liquids the angle of contact is larger than 90°.

如果固体壁面的吸附力小于分子间的内聚力,液体将不润湿表面,管内液面将下降,如图1.5(c)所示。与玻璃管接触的水银就是不润湿液体的典型例子。对于不润湿液体,接触角大于90°。

Surface tension effects play an important role in many fluid mechanics problems, such as the movement of liquids through soil and other porous media, flow of thin films, formation of drops and bubbles. However, in some fluid mechanics problems, surface tension is not important when inertial, gravitational and viscous forces are much more dominant.

表面张力效应在许多流体力学问题中具有重要作用,如:土壤和多孔介质中的流体流动、薄膜流动、液滴和气泡的形成。然而,在有些流体力学问题中,惯性力、重力和黏性力占主导地位,表面张力并不重要。

1.4.3 Viscosity

1.4.3 黏度

Density is not sufficient to uniquely characterize behavior of fluids because two fluids (such as water and oil) may have approximately the same value of density but behave quite differently when flowing. This means that some additional property is needed to describe the fluidity of the fluid.

两种密度相近的流体(如:水和油)流动起来却相差很大,可见密度不足以表征流体的行为。这意味着需要一些额外的属性来描述流体的流动性。

To determine this property, a hypothetical experiment can be considered by placing a fluid between two infinite parallel plates, as shown in Fig. 1.6. When the force F is applied to the upper plate, it will move continuously with a velocity, U. The fluid in contact with the upper plate moves with the plate at velocity, U. The velocity of the fluid in contact with the bottom fixed plate is zero. The fluid between the two plates moves with velocity $u = u(y)$ that varies linearly,

为了确定这个属性,假想一个如图1.6所示的实验,两块无限大平行平板间充满流体,下板固定,当力F作用在上板上时,上板以速度U向右运动,与上板接触的流体以同样的速度运动,与下板接触的流体速度为零。两板间流体以速度$u=u(y)$运动,速度线性变化,$u=Uy/b$,如图1.6所示。

$u = Uy/b$, as illustrated in Fig. 1.6. Thus, in this particular case a velocity gradient, du/dy, is U/b, but in more complex flow situations this would not be true. The velocity of the fluid in contact with the solid wall is equal to the velocity of the wall, and this is usually referred to as the no-slip condition. All fluids, both liquids and gases, satisfy this condition.

因此，这种情况下，速度梯度 $du/dy = U/b$，但在更复杂的流动情况下，速度分布并非如此。与固壁接触的流体速度等于固壁速度，这通常称为无滑移条件，所有流体均满足该条件。

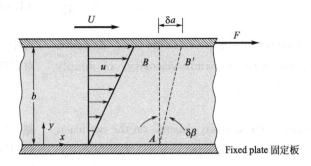

Fig. 1.6 Behavior of a fluid placed between two parallel plates
图1.6 两平行平板间的流体行为

In a small time increment, δt, an imaginary vertical line AB would rotate to line AB', so that

在 δt 时间内，假想的垂直线 AB 旋转至直线 AB'，所以有

$$\tan\delta\beta \approx \delta\beta = \frac{\delta a}{b}$$

Since $\delta a \approx U\delta t$, it follows that

因为 $\delta a \approx U\delta t$，则有

$$\delta\beta = \frac{U\delta t}{b}$$

It can be found that $\delta\beta$ is a function not only of the force F (which governs U) but also of time. Thus, the rate of shearing strain is defined as

可见，$\delta\beta$ 不仅是力 F（控制着速度 U）的函数，而且是时间的函数，因此，切应变速率定义为

$$\dot{\gamma} = \lim_{\delta t \to 0} \frac{\delta\beta}{\delta t}$$

which is equal to

它等于

$$\dot{\gamma} = \frac{U}{b} = \frac{du}{dy}$$

As the shearing stress, τ, is increased by increasing F ($\tau = F/A$), the rate of shearing strain is increased in direct proportion

由于切应力 τ 随着 F 的增大而增大（$\tau = F/A$），切应变速率成比例增加

$$\tau \propto \dot{\gamma} \quad \text{or} \quad \tau \propto \frac{du}{dy}$$

This result indicates that for common fluids the shearing stress and rate of shearing strain (velocity gradient) hold a relationship as

该式表明，对于常见流体，切应力和切应变速率（速度梯度）之间存在如下关系

$$\tau = \mu \frac{du}{dy} \quad (1.19)$$

where the constant of proportionality, μ, is called the absolute viscosity, dynamic viscosity, or simply the viscosity of the fluid.

式中，比例常数 μ 称作流体的绝对黏度、动力黏度或者黏度。

The actual value of the viscosity depends on the particular fluid and temperature as illustrated in Fig. 1.7. Fluids for which the shearing stress is linearly related to the rate of shearing strain (also referred to as rate of angular deformation) are designated as Newtonian fluids. Most common fluids are Newtonian.

黏度的实际值取决于特定的流体和温度，如图1.7所示。切应力与切应变速率（也称为角变形率）呈线性关系的流体称作牛顿流体，大部分常见流体为牛顿流体。

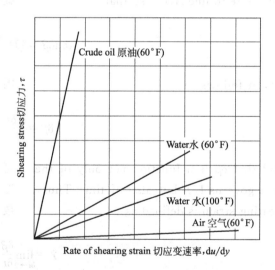

Fig. 1.7　Variation of shearing stress with rate of shearing strain for common fluids
图1.7　常见流体切应力随剪切应变率的变化

Fluids for which the shearing stress is not linearly related to the rate of shearing strain are designated as non-Newtonian fluids. Although there are a variety of types of non-Newtonian fluids, the most common are shown in Fig. 1.8. The slope of the shearing stress vs.

切应力与切应变速率成非线性关系的流体称作非牛顿流体。尽管有多种类型的非牛顿流体，常见的非牛顿流体如图1.8所示。切应力与切应变速率的斜率称为表观

rate of shearing strain graph is denoted as the apparent viscosity, μ_{ap}. For Newtonian fluids the apparent viscosity is the same as the viscosity and is independent of shear rate.

黏度，μ_{ap}。牛顿流体的表观黏度即为动力黏度，与剪切速率无关。

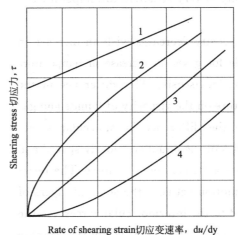

Fig. 1.8　Variation of shearing stress with rate of shearing strain for different types of fluids
图 1.8　不同类型流体的切应力随切应变速率的变化
1—Bingham plastic Bingham 塑性体；2—Shear thinning fluids 剪切变稀流体；
3—Newtonian fluids 牛顿流体；4—Shear thickening fluids 剪切增稠流体

For shear thinning fluids the apparent viscosity decreases with increasing shear rate. Common examples of this type of fluid include colloidal suspensions and polymer solutions.

剪切变稀流体的表观黏度随着剪切速率的增加而减小，常见的有：胶体悬浮液和聚合物溶液。

For shear thickening fluids the apparent viscosity increases with increasing shear rate. Common examples of this type of fluid include water-corn starch mixture and water-sand mixture.

剪切增稠流体的表观黏度随着剪切速率的增加而增加，常见的有：水和淀粉的混合液、水和沙的混合液。

A Bingham plastic is neither a fluid nor a solid. Such material can withstand a finite shear stress without motion (therefore, it is not a fluid), but once the yield stress is exceeded it flows like a fluid. Hence, it is not a solid. Toothpaste and mayonnaise are common examples of Bingham plastic materials.

Bingham 塑性体既非流体亦非固体，这种材料可以像固体一样承受一定的切应力而不发生运动，一旦超过屈服应力就会像流体一样流动起来，常见的有牙膏、沙拉酱。

From Eq. (1.19) it can be deduced that the dimensions of viscosity are FTL^{-2}, and the unit is N·s/m². Values

由式(1.19)可推得黏度的量纲为 FTL^{-2}，单位为 N·s/m²。常见

of viscosity for several common liquids and gases are listed in Table 1.3 and Table 1.5. It can be seen in these tables that viscosity is only mildly dependent on pressure and usually the effect of pressure is neglected. However, viscosity is very sensitive to temperature.

For liquids the viscosity decreases with an increase in temperature, while for gases an increase in temperature causes an increase in viscosity. This difference in the effect of temperature on the viscosity of liquids and gases is related to the difference in molecular structure. The molecules of the liquid are closely spaced and the molecules have a strong cohesive force. The resistance to relative motion between the adjacent layers of the fluid is related to the intermolecular forces. As the temperature increases, these cohesive forces decrease and the corresponding resistance to motion decreases. Since viscosity is an index of this resistance, it follows that the viscosity is reduced by an increase in temperature. However, for gases the molecules are widely spaced and intermolecular forces are negligible. In this case resistance to relative motion arises due to the exchange of momentum of gas molecules between adjacent layers. As the temperature increases, the random molecular activity increases, the momentum exchange increases, and the corresponding viscosity increases as well.

The kinematic viscosity, which is denoted with the Greek symbol ν, is the ratio of dynamic viscosity to density

$$\nu = \frac{\mu}{\rho}$$

The dimensions of kinematic viscosity are L^2T^{-1} and the unit is m^2/s. Values of kinematic viscosity for some common liquids and gases are given in Table 1.3 and Table 1.5.

液体和气体的黏度如表1.3和表1.5所列。由表可见，黏度受压力的影响较小，压力对黏度的作用通常被忽略，但黏度对温度非常敏感。

液体的黏度随温度的升高而减小，而气体的黏度随温度的升高而增大。温度对液体和气体黏度的影响不同，这与分子结构的差异有关。液体分子间距离较小，分子间有很强的内聚力，流体相邻层之间相对运动的阻力与分子间作用力有关。随着温度的升高，黏着力下降，相应的运动阻力下降。因为黏度是阻力的一个指标，所以随着温度的升高，黏度下降。然而，对于气体，分子间距离较大，分子间作用力可以忽略。在这种情况下，由于相邻层间气体分子的动量交换而产生相对运动阻力。随着温度升高，分子的随机运动增强，动量交换增大，相应的黏度也就随之增大。

运动黏度为动力黏度与密度之比，用希腊符号ν表示

运动黏度的量纲为L^2T^{-1}，单位为m^2/s。常见液体和气体的运动粘度值见表1.3和表1.5。

Exercises

1.1 Determine the dimensions for (a) the product of mass times velocity, (b) kinetic energy divided by area, and (c) the product of force times volume.

1.2 If F is a force and x is a length, what are the dimensions of dF/dx, d^3F/dx^3 and $\int F\,dx$?

1.3 A spherical particle moves slowly in a liquid. The force that is exerted on the particle can be expresses as $F = 3\pi\mu Dv$, where μ is the dynamic viscosity, D is the particle diameter, and v is the particle velocity. What are the dimensions of the constant, 3π?

1.4 The specific weight of a certain liquid is 8000N/m^3. Determine its density and specific gravity.

1.5 The gap between two parallel plates is filled with crude oil having a viscosity of $0.0035\text{Pa}\cdot\text{s}$. The bottom plate is fixed and the upper plate moves when a force F is applied. If the distance between the two plates is 2mm and the effective area of the upper plate is 1m^2, what value of F is required to translate the plate with a velocity of 1m/s?

1.6 A 20-kg, 0.2m × 0.2m bottom edge, 0.3-m-tall cube slider slides down a ramp with a constant velocity as shown in Fig. 1.9. The uniform-thickness oil layer on the ramp has a viscosity of $0.00729\text{Pa}\cdot\text{s}$. The angle of the ramp is 30°. Determine the velocity of the slider.

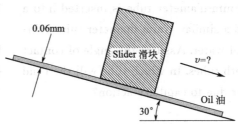

Fig. 1.9 Figure for the exercise 1.6

1.7 A piston having a diameter of 0.14m, a length of 0.25m, and a mass of 0.25kg slides downward with a velocity v through a vertical pipe. The oil film thickness between the piston and the pipe wall is 0.05mm. If the oil viscosity is 0.004Pa·s and the velocity distribution in the gap is linear, determine the velocity of the piston.

1.7 垂直圆管内有一直径为0.14m，长度为0.25m，质量为0.25kg的活塞，以速度 v 下滑。活塞与管壁间的油膜厚度为0.05mm。如果油的黏度为0.004Pa·s，间隙内的速度分布是线性的，试确定活塞的下滑速度。

1.8 A 0.3-m-diameter circular plate is placed over a fixed bottom plate with a 2.5mm gap between the two plates filled with water as shown in Fig. 1.10. Assume that the velocity distribution in the gap is linear and that the shear stress on the edge of the rotating plate is negligible. Determine the torque required to rotate the circular plate at 150r/min.

1.8 某直径为0.3m的圆盘放置在固定的底盘上，两盘的间隙为2.5mm，其间充满水，如图1.10所示。假设间隙内的速度线性分布，圆盘边缘的切应力忽略不计。试确定让圆盘以150r/min的速度旋转所需的扭矩。

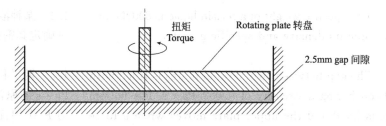

Fig. 1.10 Figure for the exercise 1.8
图 1.10 习题 1.8 图

1.9 There are two parallel disks of diameter d and the distance between the two disks is δ, dynamic viscosity of fluid in the distance is μ. If the lower disk is fixed, and the upper disk rotates in a constant angular velocity ω, prove the torque M is

1.9 两平行圆盘直径为 d，两盘间隙为 δ，盘间液体动力黏度为 μ。如果下盘固定，上盘以角速度 ω 旋转，试证明旋转所需扭矩 M 为

$$M = \frac{\pi \mu \omega d^4}{32\delta}$$

1.10 An open 2-mm-diameter tube is inserted into a cup of alcohol and a similar 4-mm-diameter tube is inserted into a cup of water. Assume the angle of contact is the same for both tubes. In which tube will the fluid column rise higher due to capillary action?

1.10 一根直径为2mm的圆管插入酒精中，另一根直径为4mm的圆管插入水中。假设两根管子的接触角相同，请问哪根管内的液柱由于毛细作用上升得更高？

Chapter 2　Fluid Statics
第2章　流体静力学

The fluid concerned in this chapter is either at rest or moving in such a manner that there is no relative motion between adjacent molecules. In both cases there are no shearing stresses in the fluid, and the only forces are due to the pressure. The absence of shearing stresses greatly simplifies the analysis to many practical problems.

本章所涉及的是静止流体或分子间没有相对运动时的流动流体，这两种情况下流体均不受切应力的作用，只受到由于压力而产生的力。没有切应力存在使许多实际问题的分析大为简化。

2.1　Pressure at a Point

2.1　某一点的压力

To analyze variation of the pressure at a point with the direction, a small triangular wedge of fluid is extracted from a fluid mass, as illustrated in Fig. 2.1. Since there are no shearing stresses in the fluid, the only external forces exerting on the wedge are resulted from the pressure and the weight. For simplicity the forces in the x direction are not shown, and the z axis is taken as the vertical axis. The weight acts in the negative z direction. To make the analysis as general as possible, the fluid element is allowed to have accelerated motion. The assumption of no shearing stress will still be valid so long as there is no relative motion between adjacent elements.

为了分析某一点上的压力随方向的变化，在流体中取任一楔形流体单元进行受力分析，如图 2.1 所示。因为没有切应力的作用，作用在楔形单元上的外力只有压力和重力。为简便起见，x 方向上的力未标出，z 轴所示为竖直方向，重力作用在负 z 方向上。为了使分析具有普遍意义，允许微元体具有加速度。只要相邻单元间没有相对运动，流体中不存在切应力的假设仍然成立。

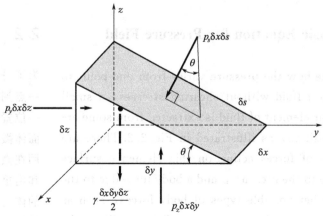

Fig. 2.1　Forces on a wedge-shaped fluid element
图 2.1　作用在楔形流体单元上的力

According to Newton's second law, the equations of motion in the y and z directions are, respectively

$$\sum F_y = p_y \delta x \delta z - p_s \delta x \delta s \sin\theta = \rho \frac{\delta x \delta y \delta z}{2} a_y$$

$$\sum F_z = p_z \delta x \delta y - p_s \delta x \delta s \cos\theta - \gamma \frac{\delta x \delta y \delta z}{2} = \rho \frac{\delta x \delta y \delta z}{2} a_z$$

where p_s, p_y and p_z are the average pressures on the faces, a_y and a_z are the accelerations in y and z directions, and γ and ρ are the fluid specific weight and density. The geometry of the wedge gives

$$\delta y = \delta s \cos\theta \qquad \delta z = \delta s \sin\theta$$

so that the equations of motion can be rewritten as

$$p_y - p_s = \rho a_y \frac{\delta y}{2}$$

$$p_z - p_s = (\rho a_z + \gamma)\frac{\delta z}{2}$$

Since we are interested in the pressure at a point, we take the limit as δx, δy and δz approach zero while maintain the angle, and it follows that

$$p_y = p_s \qquad p_z = p_s$$

Since the angle is arbitrarily chosen, it can be concluded that the pressure at a point in a fluid at rest or in motion is independent of direction as long as there are no shearing stresses. This result is called Pascal's law.

2.2 Basic Equation for Pressure Field

To analyze how the pressure varies from one point to another in a fluid without shearing stresses, a small rectangular element of fluid is extracted from some arbitrary position, as illustrated in Fig. 2.2. There are two types of forces acting on this element: surface forces due to the pressure, and a body force due to the weight. Other possible types of body forces, such as those due to electric and magnetic fields, will not be considered.

式中，p_s、p_y 和 p_z 为各面上的平均压力；a_y 和 a_z 为 y、z 方向的加速度；γ 和 ρ 为流体的重度和密度。由几何关系可得

所以运动方程可以写为

由于我们关心的是某一点的压力，令 δx、δy 和 δz 趋近于零，而角度保持不变，由此得出

因为角度是任意选取的，所以，只要没有切应力的存在，静止或运动流体中的压力与方向无关，这一结果被称为帕斯卡定律。

2.2 压力场基本方程

为了分析无切应力流体中压力从一点到另一点的变化规律，在任一位置选取一个如图2.2所示的六面体微元进行分析。有两类力作用在微元上：由压力引起的面力和由重力引起的体力。其他类型的体力（如：由电场和磁场引起的体力）不予考虑。

If the pressure at the element center is p, the average pressure on the six surfaces can be expressed in terms of p and its derivatives as shown in Fig. 2.2. The pressure is expressed in the form of Taylor series, and the higher order terms are neglected when δx, δy and δz approach zero. For simplicity the surface forces in the x direction are not shown. The resultant surface force in the y direction is

设微元中心的压力为 p，则六个面上的平均压力可以用 p 和它的导数来表示，如图 2.2 所示。压力采用泰勒级数的形式表示，当 δx、δy 和 δz 趋近于零，忽略高阶项。为简便起见，x 方向上的面力未标出，y 方向上的表面力的合力为

$$\delta F_y = \left(p - \frac{\partial p}{\partial y}\frac{\delta y}{2}\right)\delta x \delta z - \left(p + \frac{\partial p}{\partial y}\frac{\delta y}{2}\right)\delta x \delta z$$

$$\delta F_y = -\frac{\partial p}{\partial y}\delta x \delta y \delta z$$

Similarly, the resultant surface forces in the x and z directions are, respectively

类似地，x 和 z 方向上的表面力的合力分别为

$$\delta F_x = -\frac{\partial p}{\partial x}\delta x \delta y \delta z \qquad \delta F_z = -\frac{\partial p}{\partial z}\delta x \delta y \delta z$$

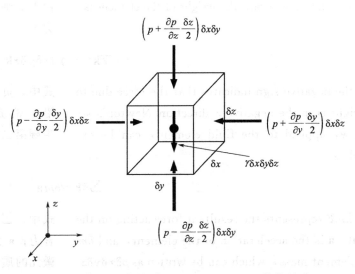

Fig. 2.2 Surface and body forces acting on a small fluid element
图 2.2 流体微元上的面力和体力

The resultant surface force acting on the element can be expressed in vector form as

作用在微元上的面力的合力采用矢量的形式可表达为

$$\delta \mathbf{F}_s = \delta F_x \mathbf{i} + \delta F_y \mathbf{j} + \delta F_z \mathbf{k}$$

or

或

$$\delta \mathbf{F}_s = -\left(\frac{\partial p}{\partial x}\mathbf{i} + \frac{\partial p}{\partial y}\mathbf{j} + \frac{\partial p}{\partial z}\mathbf{k}\right)\delta x \delta y \delta z \qquad (2.1)$$

where **i**, **j** and **k** are the unit vectors along the coordinate axes. The group of terms in parentheses in Eq. (2.1) represents the pressure gradient in vector form and can be written as

式中，**i**、**j** 和 **k** 为沿着坐标轴方向的单位矢量。式（2.1）括号中的项表示压力梯度的矢量形式，可以写成

$$\frac{\partial p}{\partial x}\mathbf{i}+\frac{\partial p}{\partial y}\mathbf{j}+\frac{\partial p}{\partial z}\mathbf{k}=\nabla p$$

where 其中

$$\nabla(\ )=\frac{\partial(\ )}{\partial x}\mathbf{i}+\frac{\partial(\ )}{\partial y}\mathbf{j}+\frac{\partial(\ )}{\partial z}\mathbf{k}$$

and the symbol ∇ presents the gradient and is commonly called Hamiltonian operator. Thus, the resultant surface force per unit volume can be expressed as

式中，∇ 表示梯度，称作哈密尔顿算子。因此，单位体积所受面力的合力可以表示为

$$\frac{\delta \mathbf{F}_s}{\delta x \delta y \delta z}=-\nabla p$$

Since the z axis is vertical, the weight of the element is

由于 z 轴在垂直方向上，微元的重力为

$$-\delta W \mathbf{k}=-\gamma \delta x \delta y \delta z \mathbf{k}$$

where the negative sign indicates that the force due to the weight is in the negative z direction. Newton's second law, applied to the fluid element, can be expressed as

式中，负号表示由重力引起的力沿 z 轴负向。应用于流体微元，牛顿第二定律可以表示为

$$\sum \delta \mathbf{F}=\delta m \mathbf{a}$$

where $\sum \delta \mathbf{F}$ represents the resultant force acting on the element, **a** is the acceleration of the element, and δm is the element mass, which can be written as $\rho \delta x \delta y \delta z$. It follows that

式中，$\sum \delta \mathbf{F}$ 表示作用在微元上的合力；**a** 为微元的加速度；δm 为微元的质量，可以写成 $\rho \delta x \delta y \delta z$。因此

$$\sum \delta \mathbf{F}=\delta \mathbf{F}_s-\delta W \mathbf{k}=\delta m \mathbf{a}$$
$$-\nabla p \delta x \delta y \delta z-\gamma \delta x \delta y \delta z \mathbf{k}=\rho \delta x \delta y \delta z \mathbf{a}$$

and, therefore 因此有

$$-\nabla p-\gamma \mathbf{k}=\rho \mathbf{a} \quad (2.2)$$

Equation(2.2) is the general equation of motion for a fluid in which there are no shearing stresses.

式（2.2）为无切应力流体运动方程的一般形式。

2.3 Pressure Variation in a Fluid at Rest

For a fluid at rest ($\mathbf{a}=0$), Eq. (2.2) reduces to

$$\nabla p + \gamma \mathbf{k} = 0$$

or in component form

$$\frac{\partial p}{\partial x}=0 \quad \frac{\partial p}{\partial y}=0 \quad \frac{\partial p}{\partial z}=-\gamma \qquad (2.3)$$

The above equations show that the pressure is independent of x and y. Thus, the pressure does not change from point to point in any plane parallel to the x-y plane. Since p depends only on z, the third of Eq. (2.3) can be written as the ordinary differential equation

$$\frac{\mathrm{d}p}{\mathrm{d}z}=-\gamma \qquad (2.4)$$

Equation (2.4) is the fundamental equation for fluids at rest and can be used to determine how pressure changes with elevation. This equation indicates that the pressure gradient in the vertical direction is negative; that is, the pressure decreases as the elevation increases. γ is not always a constant. Thus, it is valid for liquids with constant specific weight, as well as gases whose specific weight may vary with elevation. However, before integrating Eq. (2.4) the relationship between the specific weight and z must be determined.

2.3.1 Incompressible Fluid

Since the specific weight is equal to the product of fluid density and acceleration of gravity ($\gamma = \rho g$), changes in ρ or g may cause changes in γ. For most engineering problems the variation in g is negligible, so the main concern is the possible variation in the fluid density. For liquids the variation in density is usually negligible, so that the specific weight is assumed to be constant. For this instance, Eq. (2.4) can be directly integrated

2.3 静止流体中的压力变化

对于静止的流体（$\mathbf{a}=0$），式(2.2)简化为

表示为分量的形式

上式表明，压力与 x 和 y 无关。因此，压力在任何平行于 x-y 面的平面上不发生变化。因为压力只与 z 有关，式(2.3)中的第三式可以写成常微分形式

式(2.4)为静止流体的基本方程，可用于确定压力在高度方向上的变化。该式表明，压力梯度在垂直方向上是负的，也就是说，随着高度的增加，压力降低。式中重度不一定为常数，因此，对于重度不变的液体和重度随高度变化的气体，该式都是成立的。在对式(2.4)进行积分前，必须确定重度和 z 之间的关系。

2.3.1 不可压缩流体

由于重度等于密度和重力加速度的乘积（$\gamma=\rho g$），ρ 或 g 的变化都可能引起 γ 的变化。对于大多数工程问题，g 的变化可以忽略不计，所以主要关注的是流体密度的变化。对于液体，密度的变化通常可以忽略，所以通常假设液体重度为常数，对于这种情况，式(2.4)可以直接积分

$$\int_{p_1}^{p_2} dp = -\gamma \int_{z_1}^{z_2} dz$$

to yield 计算得

$$p_2 - p_1 = -\gamma(z_2 - z_1)$$

or 或

$$p_1 - p_2 = \gamma(z_2 - z_1) \quad (2.5)$$

where p_1 and p_2 are pressures at the vertical elevations as is shown in Fig. 2.3.

式中，p_1 和 p_2 为图 2.3 所示的垂直高度上的压力。

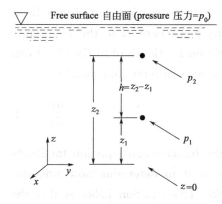

Fig. 2.3　Pressure variation in a fluid at rest
图 2.3　静止流体中的压力变化

Equation (2.5) can be written as 式(2.5) 可写为

$$p_1 - p_2 = \gamma h \quad (2.6)$$

or 或

$$p_1 = p_2 + \gamma h \quad (2.7)$$

Equation (2.7) shows that in an incompressible fluid at rest the pressure varies linearly with depth.

式(2.7) 表明，在不可压缩静止流体中，压力随深度线性变化。

The free surface shown in Fig. 2.3 is usually used as a reference plane. The reference pressure would frequently be atmospheric pressure, and thus if $p_2 = p_0$ in Eq. (2.7) it follows that the pressure p at any depth h below the free surface is given by the following equation

通常将如图 2.3 所示的自由面作为参考面，参考压力往往为大气压。因此，如果令式(2.7) 中的 $p_2 = p_0$，则自由面下任意深度 h 处的压力 p 由下式给出

$$p = p_0 + \gamma h \quad (2.8)$$

It can be found from Eq. (2.7) or Eq. (2.8) that the pressure in a homogeneous incompressible fluid at rest depends on the depth of the fluid relative to some reference plane, and it is not influenced by the size or shape of the container in which the fluid is held. Thus, in Fig. 2.4 the pressure is the same at the bottom of the containers. The pressure depends only on the depth, h, the surface pressure, p_0, and the specific weight, γ.

由式(2.7) 或式(2.8) 可知，匀质不可压缩静止流体中的压力取决于流体相对于参考面的深度，而不受容器尺寸和形状的影响。因此，图 2.4 中各容器底部的压力均相同，压力的大小只取决于深度 h、表面压力 p_0 和液体的重度 γ。

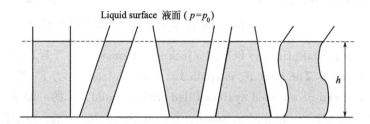

Fig. 2.4 Pressure in a container of arbitrary shape

图 2.4 任意形状容器内的压力

【EXAMPLE 2.1】 The mixture of water and gasoline is shown in Fig. 2.5. If the specific gravity of the gasoline is $SG = 0.68$, determine the pressure in the unit of Pa at the gasoline-water interface and at the bottom of the tank.

【例题 2.1】 汽油和水的混合物如图 2.5 所示，如果汽油的比重为 0.68，采用 Pa 作为压力单位，计算油水交界面及箱底的压力。

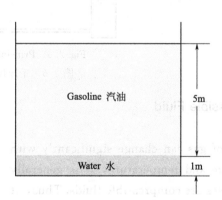

Fig. 2.5 Figure for the example 2.1

图 2.5 例题 2.1 图

SOLUTION Since the liquids are at rest, the pressure variation can be calculated by

解 由于两种液体均处于静止状态，压力可由下式计算

$$p = \gamma h + p_0$$

where p_0 is the pressure at the free surface of the gasoline. The pressure at the interface is

式中，p_0 表示汽油自由面上的压力。交界面上的压力为

$$p_1 = SG\gamma_{H_2O}h + p_0 = 0.68 \times 9789 \times 5 + p_0 = 33282.6 + p_0 \text{ (Pa)}$$

If the pressure is relative to atmospheric pressure (gage pressure), it follows that $p_0 = 0$, and therefore

如果所求压力为相对于大气压的压力（表压），则 $p_0 = 0$，因此

$$p_1 = 33282.6 \text{ (Pa)}$$

Apply the same method to determine the pressure at the tank bottom; that is

采用同样的方法确定箱底的压力，即

$$p_2 = \gamma_{H_2O}h_{H_2O} + p_1 = 9789 \times 1 + 33282.6 = 43071.6 \text{ (Pa)}$$

The fundamental principle behind a jack is demonstrated in Fig. 2.6. The force F_1 is applied at a piston located at one end of a closed system filled with a liquid, and the force F_2 is applied at another piston. Since the pressure p acting on the faces of both pistons is the same, it follows that $F_2 = (A_2/A_1)F_1$; that is, a small force applied at the smaller piston can generate a large force at the larger piston.

千斤顶的基本原理如图 2.6 所示。力 F_1 施加在充满液体的封闭系统的一端活塞上，力 F_2 施加在另一端，由于作用在两个活塞面上的压力 p 相同，则有 $F_2 = (A_2/A_1)F_1$，即，作用在小活塞上的较小的力可以在大活塞上产生较大的力。

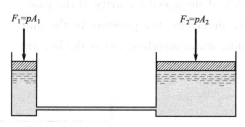

Fig. 2.6 Principle behind a jack

图 2.6 千斤顶的基本原理

2.3.2 Compressible Fluid

2.3.2 可压缩流体

Since the density of gas can change significantly with changes in pressure and temperature, it is generally considered that gases are compressible fluids. Thus, it is necessary to know the variational regularity of γ before integrating Eq. (2.4). Since the specific weights of gases are comparatively small, it follows from Eq. (2.4) that the pressure gradient in the vertical direction is correspondingly small. This means the effect of elevation changes on the pressure can be neglected if the distances are small.

由于气体的密度会随压力和温度的变化发生明显的改变，通常认为气体是可压缩流体。因此，在对式(2.4)进行积分之前必须知道重度 γ 的变化规律。由于气体的重度相对较小，由式(2.4)可知，竖直方向上的压力梯度也相应较小，这意味着如果距离很小，高度变化对压力的影响可以忽略。

If the variations in heights are large, the variation in the specific weight should be considered. As described in chapter 1, the equation of state for an ideal gas is

$$p = \rho R T$$

where p is the absolute pressure, R is the gas constant, and T is the thermodynamic temperature. Combined with Eq. (2.4), we have

$$\frac{\mathrm{d}p}{\mathrm{d}z} = -\frac{gp}{RT}$$

Then by separating variables

$$\int_{p_1}^{p_2} \frac{\mathrm{d}p}{p} = \ln \frac{p_2}{p_1} = -\frac{g}{R} \int_{z_1}^{z_2} \frac{\mathrm{d}z}{T} \qquad (2.9)$$

where g and R are assumed to be constant in the elevation range from z_1 to z_2. Although the acceleration of gravity, g, does vary with elevation, the variation is very small, and g is usually assumed constant.

The variation regularity of temperature with elevation must be specified for the integration. For example, if the temperature is assumed to be T_0 over the range z_1 to z_2, it follows from Eq. (2.9) that

$$p_2 = p_1 \exp\left[-\frac{g(z_2 - z_1)}{RT_0}\right] \qquad (2.10)$$

This equation provides the relationship between pressure and elevation for an isothermal layer. For nonisothermal conditions, if the temperature-elevation relationship is known, a similar procedure can be followed.

2.4 Standard Atmosphere

The concept of a standard atmosphere was first proposed in the 1920s. The currently accepted standard atmosphere is the so-called U.S. standard atmosphere, which is the annual mean value of the earth's atmosphere in the middle latitudes. Several important properties for standard atmospheric conditions at sea level are listed in Table 2.1. Figure 2.7 shows the variation of temperature with altitude for the U.S. standard atmosphere. It can be seen in this figure that the temperature decreases with altitude in the troposphere, then becomes constant in the stratosphere, and subsequently starts to increase in the next layer.

美国标准大气温度随高度的变化。由图可见，对流层中温度随着高度的增加而降低，平流层中温度基本不变，接着随着高度的增加，温度又开始上升。

Table 2.1　Properties of U. S. standard atmosphere at sea level
表2.1　美国海平面标准大气压的特性

Property 特性	Value 数值
Temperature 温度,T	288.15 K(15℃)
Pressure 压力,p	101.33kPa(abs)
Density 密度,ρ	1.225kg/m³
Specific weight 重度,γ	12.014N/m³
Viscosity 黏度,μ	1.789×10^{-5} N·s/m²

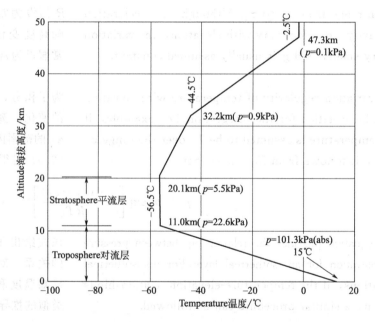

Fig. 2.7　Variation of temperature with altitude
图2.7　温度随高度的变化

Since the temperature variation is represented by a series of linear segments, it is possible to integrate Eq. (2.9) to obtain the corresponding pressure variation. For example, in the troposphere the temperature variation can be expressed as

由于温度变化由一系列线性线段表示，对式(2.9)积分即可得到相应的压力变化。例如：在对流层中温度随高度的变化可表示为

$$T = T_s - \alpha z \quad (2.11)$$

where T_s is the temperature at sea level and α is the rate of change of temperature with elevation. For the standard atmosphere in the troposphere, $\alpha = 0.0065 \text{K/m}$.

Equation (2.11) used with Eq. (2.9) yields

$$p = p_s \left(1 - \frac{\alpha z}{T_s}\right)^{g/R\alpha} \quad (2.12)$$

where p_s is the absolute pressure at seal level ($z=0$). With p_s, T_s and g obtained from Table 2.1, and with the gas constant $R = 286.9 \text{J/kg} \cdot \text{K}$, the pressure variation throughout the troposphere can be determined from Eq. (2.12). This calculation shows that the temperature is $-56.5°C$ and the absolute pressure is 23kPa at the outer edge of the troposphere. It is to be noted that jetliners cruise at approximately this altitude. Pressures at other altitudes are shown in Fig. 2.7.

2.5 Buoyancy and Stability

2.5.1 Archimedes' Principle

When a stationary body is completely submerged in a fluid or floating, the resultant fluid force acting on the body is called the buoyant force. A body of arbitrary shape having a volume V is immersed in a fluid as illustrated in Fig. 2.8(a). The body is enclosed in a parallelepiped, and a free-body diagram of the parallelepiped with the body removed is shown in Fig. 2.8(b). The forces F_1, F_2, F_3 and F_4 are simply the forces exerted on the plane surfaces of the parallelepiped (for simplicity the forces in the x direction are not shown), W is the weight of the shaded fluid volume (parallelepiped minus body), and F_B is the force the body is exerting on the fluid. The forces on the parallel surfaces, such as F_3 and F_4, are all equal and cancel, so only the equilibrium equation in the z direction needs to be considered and can be expressed as

$$F_B = F_2 - F_1 - W \quad (2.13)$$

| If the specific weight of the fluid is constant, then | 如果流体的重度为常数，那么 |

$$F_2 - F_1 = \gamma(h_2 - h_1)A_0$$

| where A_0 is the horizontal area of the upper (or lower) surface of the parallelepiped, and Eq. (2.13) can be written as | 式中，A_0 是平行六面体上表面或下表面的面积，式（2.13）可以写为 |

$$F_B = \gamma(h_2 - h_1)A_0 - \gamma[(h_2 - h_1)A_0 - V]$$

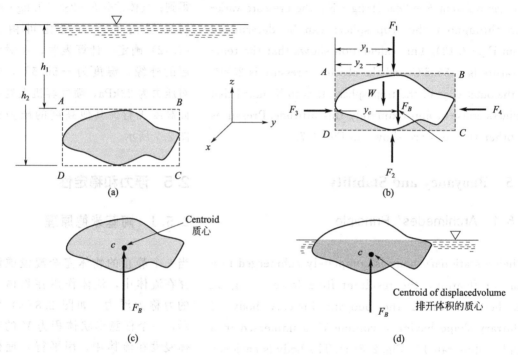

Fig. 2.8　Buoyant force on submerged and floating bodies

图 2.8　浸没和漂浮物体上的浮力

| The expression for the buoyant force can be obtained by simplifying the equation mentioned above | 通过简化上式，可以得到浮力的表达式为 |

$$F_B = \gamma V \quad (2.14)$$

| where γ is the specific weight of the fluid and V is the volume of the body. The direction of the buoyant force is opposite to that shown on the free-body diagram. Therefore, the buoyant force is equal to the weight of the fluid displaced by the body and is directed vertically upward. This is the so-called Archimedes' principle. | 式中，γ 为流体的重度；V 为物体的体积。浮力的方向与受力图中的方向相反。因此，浮力的大小为排开液体的重量，方向垂直向上，这就是所谓的阿基米德原理。 |

The location of the buoyant force can be determined by summing moments of the forces shown in Fig. 2.8(b) with respect to an axis. For example, by summing moments about an axis perpendicular to the y-z plane through point A it can be obtained that

$$F_B y_c = F_2 y_1 - F_1 y_1 - W y_2$$

and using volume instead of the various forces

$$V y_c = V_T y_1 - (V_T - V) y_2 \quad (2.15)$$

where V_T is the total volume $(h_2 - h_1) A_0$. The left-hand side of Eq. (2.15) is the first moment of the displaced volume V with respect to the x-z plane so that y_c is equal to the y coordinate of the centroid of the volume V. In a similar manner it can be obtained that the x coordinate of the buoyant force is same with the x coordinate of the centroid. Thus, it can be concluded that the buoyant force passes through the centroid of the displaced volume as shown in Fig. 2.8(c). The point where the buoyant force acts is called the center of buoyancy.

If the density of the fluid above the liquid surface is very small compared with the liquid in which the body floats, the above results can also be applied to floating bodies as illustrated in Fig. 2.8(d). For example, the liquid is water, and the fluid above the liquid surface is air.

【EXAMPLE 2.2】 A spherical buoy has a volume of 1.6m^3, weighs 8.4kN, and is anchored to the sea floor with a cable as is shown in Fig. 2.9(a). For this condition what is the tension of the cable?

SOLUTION A free-body diagram of the buoy is shown in Fig. 2.9(b), where F_B is the buoyant force acting on the buoy, W is the weight of the buoy, and T is the tension in the cable. For equilibrium it follows that

$$T = F_B - W$$

通过求受力图 2.8(b) 中各力对某一轴的力矩之和，可以确定浮力的作用位置。例如：对通过 A 点且垂直于 y-z 平面的轴求力矩，可得

采用体积代替各力，得

式中，V_T 为平行六面体的总体积 $(h_2 - h_1) A_0$。式(2.15) 的左侧为体积 V 相对于 x-z 平面的一次矩，所以 y_c 等于质心的 y 坐标。同样的方法可得，浮力的 x 坐标与质心的 x 坐标相同。因此可以下结论，浮力通过移除体的质心，如图 2.8(c) 所示。浮力作用的点称为浮心。

如果液面上流体的密度相对于液体的密度非常小，上述所得结果同样适用于如图 2.8(d) 所示的漂浮物体，如：液体为水，液面上的流体是空气。

【例题 2.2】 一个体积为 1.6m^3、重力为 8.4kN 的浮标用绳子固定在海底，如图 2.9(a) 所示。绳子上的拉力多大？

解 浮标的受力图如图 2.9(b) 所示，F_B 为作用在浮标上的浮力；W 为浮标所受重力；T 为绳子的拉力。根据受力平衡得

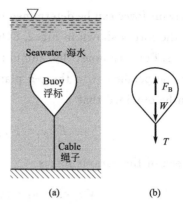

Fig. 2.9 Figure for the example 2.2

图 2.9 例题 2.2 图

From Eq. (2.14) and for seawater with $\gamma = 10.1 \text{kN/m}^3$ then

知式（2.14）且海水的重度 $\gamma = 10.1 \text{kN/m}^3$，则

$$F_B = 10.1 \times 1.6 = 16.16 \text{ (kN)}$$

The tension in the cable can now be calculated as

绳子上的拉力为

$$T = 16.16 - 8.40 = 7.76 \text{ (kN)}$$

2.5.2 Stability

2.5.2 稳定性

For a completely immersed body, if it returns to its equilibrium position when displaced, it is said to be in a stable equilibrium position. Conversely, if it moves to a new equilibrium position when displaced, it is in an unstable equilibrium position. Since the centers of buoyancy and gravity may not coincide, a small rotation can result in either a restoring or overturning couple. For example, for the completely submerged body shown in Fig. 2.10, the center of gravity (CG) is below the center of buoyancy. When the body rotates slightly, a restoring couple formed by the weight, W, and the buoyant force, F_B, will cause the body to rotate back to its original position. Thus, for this configuration the body is stable. However, for the completely submerged body illustrated in Fig. 2.11, its center of gravity is above its center of buoyancy, and therefore it is in an unstable equilibrium position. It can be noted that the body is in a stable equilibrium position as long as the center of gravity falls below the center of buoyancy.

对于完全浸没的物体，如果移动后仍能回到原先的平衡位置，则认为该物体处于稳定的平衡位置；相反，如果移动后到达新的平衡位置，则认为该物体处于不稳定的平衡位置。由于重心和浮心可能不重合，微小的旋转即可能会引起恢复或翻转的力偶。例如：图 2.10 所示为一个完全浸没的物体，重心位于浮心的下方，当发生微小的旋转时，重力 W 和浮力 F_B 形成的力偶使物体回到原先的位置，因此该物体是稳定的。然而，如图 2.11 所示，一个完全淹没的物体，其重心在浮心之上，处于不稳定的平衡位置。可见，只要重心位于浮心的下方，则物体处于稳定的平衡位置。

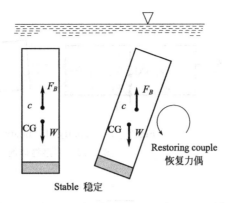

Fig. 2.10 Stability of a completely immersed body—stable configuration
图 2.10 完全浸没物体的稳定性——稳定结构

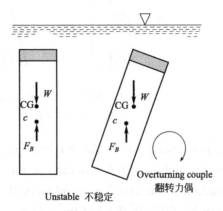

Fig. 2.11 Stability of a completely immersed body—unstable configuration
图 2.11 完全浸没物体的稳定性——不稳定结构

For a floating body, since the location of the center of buoyancy may change as the body rotates, the stability problem becomes more complicated. A relatively flat body shown in Fig. 2.12 can be stable even though the center of gravity lies above the center of buoyancy. The reasons are as follows: as the body rotates the buoyant force, F_B, shifts to pass through the centroid of the newly formed displaced volume and combines with the weight to form a couple, which can cause the body to return to its original equilibrium position. However, for the relatively slender body shown in Fig. 2.13, even a small rotational displacement can cause the buoyant force and the weight to form an overturning couple. In Fig. 2.12 and Fig. 2.13, c and c' represent the centroid of original displaced volume and the centroid of new displaced volume, respectively.

对于漂浮的物体，由于物体的旋转将使浮心的位置发生改变，稳定性问题变得更为复杂。如图 2.12 所示的相对扁平的物体，即使重心在浮心的上方，也能保持稳定。原因如下：物体旋转后浮力通过新形成的排开体积的质心，和重力一起形成力矩，能够使物体回到原先的平衡位置。然而，对于如图 2.13 所示的相对细长的物体，即便很小的旋转也能使浮力和重力形成翻转力矩。在图 2.12 和图 2.13 中，c 和 c' 分别表示原先排开体积的质心和新形成的排开体积的质心。

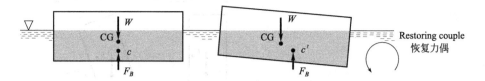

Fig. 2.12 Stability of a floating body—stable configuration
图 2.12 漂浮物体的稳定性——稳定结构

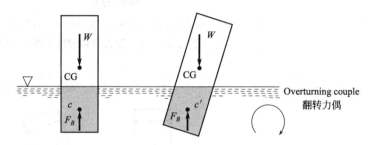

Fig. 2.13 Stability of a floating body—unstable configuration
图 2.13 漂浮物体的稳定性——不稳定结构

These examples show that, since the analysis is related to the geometry and weight distribution of the body, the determination of the stability of submerged or floating bodies can be difficult. If other types of external forces such as wind gusts or currents are included, the problem can be further complicated. Obviously, it is necessary to considerate the stability in the design of ships, submarines, bathyscaphes and so on.

这些例子表明，由于稳定性与物体的几何形状和重量分布有关，确定浸没或漂浮物体的稳定性很困难。如果还受到风力和水流等其他外力的作用，问题将会变得更加复杂。显然，在设计轮船、潜艇、深海潜水器等时，有必要考虑稳定性。

2.6 Measurement of Pressure

2.6 压力的测量

Pressure is a force per unit area, and the unit of pressure is N/m^2; this combination is called the pascal and written as Pa. Pressure can also be expressed as the height of a column of liquid. Then, the unit will refer to the height of the column (mm, m, etc.), and in addition, the liquid in the column must be specified (H_2O, Hg, etc.). For example, standard atmospheric pressure can be expressed as 760mmHg (abs).

压力表示单位面积上的力，单位为 N/m^2，称作帕斯卡，Pa。压力也可以用液柱的高度来表示，其单位就采用液柱高度的单位（如：毫米、米等），另外必须指定液体的种类（如：水、水银等）。例如：标准大气压可表示为 760mmHg (abs)。

The pressure can be designated as either an absolute pressure or a gage pressure. Absolute pressure is measured

压力可以表示为绝对压力或表压，绝对压力是相对于绝对零压的压力，

relative to absolute zero pressure, whereas gage pressure is measured relative to the local atmospheric pressure. Thus, a gage pressure of zero is equal to the local atmospheric pressure. Absolute pressures are always positive, but gage pressures can be either positive or negative depending on whether the pressure is above or below atmospheric pressure. The concept of these pressures is illustrated in Fig. 2.14. In this book, pressures will be assumed to be gage pressures unless specifically designated absolute. For example, 1MPa would be gage pressure, whereas 1MPa (abs) would be absolute pressure.

而表压是相对于当地大气压的压力。因此，表压为零即为当地的大气压。绝对压力总是为正，而表压可正可负，取决于压力高于大气压还是低于大气压。这些压力的概念如图 2.14 所示。本书中，除非明确指定绝对压力，使用的压力均认为是表压，例如：1MPa 指的是表压，而 1MPa（abs）指的是绝对压力。

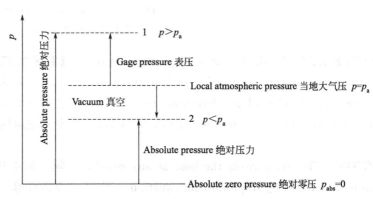

Fig. 2.14 Graphical representation of gage and absolute pressure
图 2.14 表压和绝对压力的图形表示

The mercury barometer is usually used to measure the atmospheric pressure. It consists of a glass tube closed at one end. The tube is initially filled with mercury and then turned upside down with the open end in the container of mercury as shown in Fig. 2.15. When the liquid level is in equilibrium, the weight of the mercury column plus the force due to the vapor pressure is equal to the force due to the atmospheric pressure. Thus

$$p_a = \gamma h + p_{vapor} \quad (2.16)$$

where γ is the specific weight of mercury. For most practical situations the contribution of the vapor pressure can be neglected since it is very small so that $p_a \approx \gamma h$. It is conventional to specify atmospheric pressure in terms of the height, h, in millimeters of mercury.

通常采用水银气压计进行大气压力的测量，它由一根一端封闭玻璃管组成，玻璃管最初充满水银，然后翻转使开口的一端插入水银槽中，如图 2.15 所示。当液面处于平衡位置时，水银柱的重力加上蒸气压产生的力等于大气压产生的力，因此

式中，γ 为水银的重度。对于大多数实际情况，因为水银蒸气压很小，可忽略不计，所以 $p_a \approx \gamma h$。习惯上采用毫米汞柱来表示大气压力。

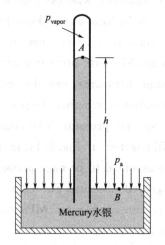

Fig. 2.15 Mercury barometer
图 2.15 水银气压计

【EXAMPLE 2.3】 A lake has an average temperature of 10℃ and a maximum depth of 50m. For an atmospheric pressure of 600mmHg, determine the absolute pressure (in pascal) at the deepest part of the lake.

SOLUTION The pressure in the lake at any depth, h, can be calculated by the following equation

$$p = \gamma h + p_0$$

where p_0 is the pressure at the lake surface. For $\gamma_{Hg} = 133 \text{kN/m}^3$

$$p_0 = 133 \times 0.6 = 79.8 \text{ (kN/m}^2\text{)}$$

$\gamma_{H_2O} = 9.804 \text{kN/m}^3$ at 10℃ and therefore

$$p = 9.804 \times 50 + 79.8 = 570 \text{ (kN/m}^2\text{)} = 570 \text{ (kPa) (abs)}$$

【例题 2.3】 一个湖的平均温度为 10℃，最大深度为 50m。若大气压力为 600mmHg，试确定湖底最深处的绝对压力（用 Pa 作为单位）。

解 湖中任意深度 h 的压力可由下式计算

$$p = \gamma h + p_0$$

式中，p_0 为湖面压力。水银的重度为 133kN/m³，则

$$p_0 = 133 \times 0.6 = 79.8 \text{ (kN/m}^2\text{)}$$

10℃下水的重度为 9.804kN/m³，则

$$p = 9.804 \times 50 + 79.8 = 570 \text{ (kN/m}^2\text{)} = 570 \text{ (kPa) (abs)}$$

2.7 Manometry

One of the methods of pressure measurement is to use a liquid column in a vertical or inclined tube to measure the pressure. Pressure measuring devices based on this method are called manometers. Three common types of manometers include the piezometer tube, the U-tube manometer, and the inclined-tube manometer.

2.7 压力测量法

压力测量的方法之一是使用垂直或倾斜管中的液柱进行压力测量，基于该方法的压力测量装置称作压力计。常见的三种压力计包括：测压管、U 形管压力计和倾斜管压力计。

2.7.1 Piezometer Tube

The simplest type of manometer consists of a vertical tube as illustrated in Fig. 2.16. The tube is open at the top, and the other end is connected to the container in which the pressure is measured. Since the fluid in the manometer is at rest, the pressure measuring principle can be described by Eq. (2.8).

2.7.1 测压管

最简单的压力计由一根垂直管构成,如图 2.16 所示,上端开口,另一端与所测压力的容器相连。由于压力计内流体处于静止状态,测压原理可用式(2.8)进行描述。

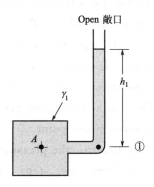

Fig. 2.16 Piezometer tube

图 2.16 测压管

The pressure p_A at point A in Fig. 2.16 can be determined according to the following relationship

图 2.16 中 A 点的压力 p_A 可以根据下列关系确定

$$p_A = p_1 = \gamma_1 h_1$$

where γ_1 is the specific weight of the liquid in the container, and h_1 is the height of the liquid column. Note that since the tube is open at the top, the pressure p_0 can be set equal to zero if the gage pressure is used. Since point ① and point A are at the same elevation, $p_A = p_1$.

式中,γ_1 为容器内液体的重度;h_1 为液柱的高度。注意:由于压力管上端开口,如使用表压,可设置 p_0 为 0。由于点①和 A 处于同一高度,故 $p_A = p_1$。

Although the piezometer tube is an accurate pressure measuring device, it has several disadvantages.
ⅰ. The pressure in the container must be greater than atmospheric pressure, otherwise air would be sucked into the system.
ⅱ. The pressure to be measured must be relatively small so the required height of the liquid column is reasonable.

尽管测压管是一种精确的测压装置,但它也有一些局限。
ⅰ. 容器内压力必须大于大气压,否则空气会被吸入。
ⅱ. 被测压力应该相对较小,以保证液柱的高度合理。

ⅲ. The fluid in the container must be a liquid rather than a gas.

2.7.2 U-Tube Manometer

To overcome the disadvantages noted previously, U-Tube Manometer is adopted as shown in Fig. 2.17. The fluid in the manometer is called the gage fluid. To determine the pressure in terms of the column heights, the pressure is calculated point by point from one end of the system to the other end by utilizing Eq. (2.8). The pressures at points A and ① are the same and it will increase by $\gamma_1 h_1$ from point ① to ②. The pressure at point ② is equal to the pressure at point ③, since the pressures at equal elevations in a continuous mass of fluid at rest must be the same. From point ③ to the open end where the pressure is zero, the pressure decreases by $\gamma_2 h_2$. In equation form these steps can be expressed as

$$p_A + \gamma_1 h_1 - \gamma_2 h_2 = 0$$

Thus, the pressure p_A can be expressed in terms of the column heights as

$$p_A = \gamma_2 h_2 - \gamma_1 h_1 \quad (2.17)$$

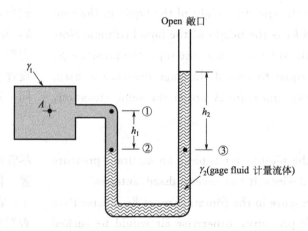

Fig. 2.17 U-tube manometer
图 2.17 U 形管压力计

A major advantage of the U-tube manometer is the fluid in the tube can be different from the fluid in the

ⅲ. 容器内的流体应为液体而非气体。

2.7.2 U形管压力计

为了克服上述局限性，可采用如图 2.17 所示的 U 形管压力计。压力计中的流体称作计量流体。为了根据液柱高度确定压力，采用式(2.8)从系统的一端到另一端逐点计算压力。点 A 和①的压力相等，从点①到点②，压力升高 $\gamma_1 h_1$。由于在相同连续介质中点②和点③的高度相同，两者压力相等。从点③到开口端（压力为0），压力降低 $\gamma_2 h_2$。上述步骤用方程的形式可表达为

因此，压力 p_A 用液柱高度的形式可以表达为

U 形管压力计的一大优点是管内流体可以与容器内流体不同。例如：

container. For example, the fluid in the container can be either a liquid or a gas. If it does contain a gas, the contribution of the gas column, $\gamma_1 h_1$, is almost always negligible so that $p_A \approx p_2$. In this instance Eq. (2.17) becomes

容器内的流体可以是液体也可以是气体。如果容器内为气体，气柱产生的压力 $\gamma_1 h_1$ 通常被忽略，所以 $p_A \approx p_2$，这样式（2.17）变为

$$p_A = \gamma_2 h_2$$

Thus, for a given pressure the height, h_2, is related to the specific weight, γ_2, of the gage fluid. If the pressure p_A is large, a heavy gage fluid should be used to maintain a reasonable column height. Alternatively, if the pressure p_A is small, a light gage fluid should be used to achieve a relatively large column height, which is easily read.

因此，对于一定的压力，高度 h_2 与计量流体的重度 γ_2 有关。如果压力 p_A 较大，应采用较重的流体以保持合理的液柱高度；如果压力 p_A 较小，应采用较轻的液体以得到相对较大的液柱高度，以便读数。

【EXAMPLE 2.4】 As shown in Fig. 2.18, a closed tank contains compressed air and oil, and a U-tube manometer using mercury is connected to the tank. $SG_{oil}=0.80$, $SG_{Hg}=13.6$, $h_1=900\text{mm}$, $h_2=150\text{mm}$, and $h_3=230\text{mm}$. Determine the pressure reading (in kPa) of the gage.

【例题 2.4】 如图 2.18 所示，一个装有压缩空气和油的密闭储罐，与水银 U 形管压力计相连。$SG_{oil}=0.80$，$SG_{Hg}=13.6$，$h_1=900\text{mm}$，$h_2=150\text{mm}$，$h_3=230\text{mm}$。试确定压力表的读数。

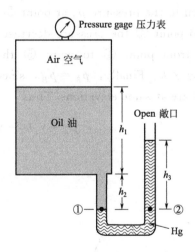

Fig. 2.18 Figure for the example 2.4
图 2.18 例题 2.4 图

SOLUTION The pressure is calculated from air-oil interface in the tank to the open end where the pressure is zero. The pressure at point ① is

解 压力从油气交界面计算到 U 形管压力计的开口端。点①处的压力为

$$p_1 = p_{air} + \gamma_{oil}(h_1 + h_2)$$

This pressure is equal to the pressure at point ②, since these two points are at the same elevation. From point ② to the open end, the pressure decreases by $\gamma_{Hg} h_3$, and at the open end the pressure is zero. Thus, the pressure can be expressed as

由于点①和②的高度相同，两位置处的压力相等。从点②到开口端，压力下降 $\gamma_{Hg} h_3$，开口端的压力为零。因此，压力可表达为

$$p_{air} + \gamma_{oil}(h_1 + h_2) - \gamma_{Hg} h_3 = 0$$

or 或

$$p_{air} + SG_{oil} \gamma_{H_2O}(h_1 + h_2) - SG_{Hg} \gamma_{H_2O} h_3 = 0$$

For the values given 代入数据，得

$$p_{air} = 13.6 \times 9802 \times 0.23 - 0.8 \times 9802 \times (0.9 + 0.15) = 22427 \text{ (Pa)}$$

The U-tube manometer is also widely used to measure the difference in pressure between two containers or two points, as shown in Fig. 2.19. The difference in pressure between point A and B can be obtained by calculating the pressure from one end to the other end. For example, at point A the pressure is p_A, which is equal to pressure p_1 at point ①. From point ① to point ② the pressure increases by $\gamma_1 h_1$. The pressure p_2 at point ② is equal to the pressure p_3 at point ③, and from point ③ to point ④ the pressure decreases by $\gamma_2 h_2$. Similarly, from point ④ to point ⑤ the pressure decreases by $\gamma_3 h_3$. Finally, $p_5 = p_B$, since point ⑤ and point B are at same elevations. Thus

U形管压力计也常用于测量两个容器或两点之间的压力差，如图2.19所示。从一端计算到另一端，可以得到点 A 和点 B 之间的压差。例如：点 A 的压力为 p_A，等于点①处的压力 p_1；从点①到点②，压力增加 $\gamma_1 h_1$；点②处的压力 p_2 等于点③处的压力 p_3；从点③到点④压力下降 $\gamma_2 h_2$，类似地，从点④到点⑤压力下降 $\gamma_3 h_3$；由于点⑤和点 B 高度相同，$p_5 = p_B$。因此

$$p_A + \gamma_1 h_1 - \gamma_2 h_2 - \gamma_3 h_3 = p_B$$

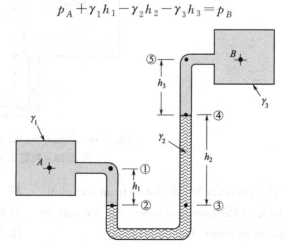

Fig. 2.19 Differential U-tube manometer
图 2.19 U型管差压计

and the pressure difference is

压差为

$$p_A - p_B = \gamma_2 h_2 + \gamma_3 h_3 - \gamma_1 h_1$$

Capillarity due to surface tension is usually not considered in the manometer. Of course, the gage fluid must be immiscible with respect to the other fluids in contact with it. To achieve accurate measurements, special attention should be paid to the change in the specific weight of fluid caused by the temperature change.

在压力计中，通常不考虑由于表面张力引起的毛细管作用。当然，计量流体必须与接触的流体不相溶。为了达到精确的测量，应该特别注意温度变化引起的流体重度的变化。

【EXAMPLE 2.5】 As illustrated in Fig. 2.20, the orifice creates a pressure drop, $p_A - p_B$, along the pipe, which is related to the flow rate through the equation $Q = K(p_A - p_B)^{1/2}$ where K is a constant. The pressure drop is measured by using a differential U-tube manometer. (a) Determine an equation for $p_A - p_B$ in terms of the specific weight of the flowing fluid, γ_1, the specific weight of the gage fluid, γ_2, and the various heights indicated. (b) For $\gamma_1 = 8.60 \text{kN/m}^3$, $\gamma_2 = 15.60 \text{kN/m}^3$, $h_1 = 0.8\text{m}$, and $h_2 = 0.6\text{m}$, what is the value of the pressure drop, $p_A - p_B$?

【例题 2.5】 如图 2.20 所示，管内流体流经孔板产生压降 $p_A - p_B$，压降与流量的关系为 $Q = K(p_A - p_B)^{1/2}$，其中 K 为常数。采用 U 形管压力计测量压降。(a) 根据 γ_1、γ_2 及相关高度，确定 $p_A - p_B$ 的表达式。(b) 如果 $\gamma_1 = 8.60 \text{kN/m}^3$，$\gamma_2 = 15.60 \text{kN/m}^3$，$h_1 = 0.8\text{m}$，$h_2 = 0.6\text{m}$，则压降 $p_A - p_B$ 为多大？

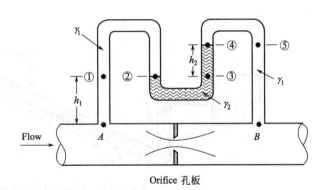

Fig. 2.20 Figure for the example 2.5
图 2.20 例题 2.5 图

SOLUTION (a) Although the fluid in the pipe is moving, the fluids in the columns of the manometer are at rest. The pressure is calculated along the path $A \rightarrow ① \rightarrow ② \rightarrow ③ \rightarrow ④ \rightarrow ⑤ \rightarrow B$. In equation form

解 (a) 虽然管内流体是流动的，但压力计内的流体是静止的。沿路径 $A \rightarrow ① \rightarrow ② \rightarrow ③ \rightarrow ④ \rightarrow ⑤ \rightarrow B$ 计算压力，可得

$$p_A - \gamma_1 h_1 - \gamma_2 h_2 + \gamma_1 (h_1 + h_2) = p_B$$

$$p_A - p_B = h_2(\gamma_2 - \gamma_1)$$

It should be noted that the pressure difference is only related to h_2 and not related to h_1. If the difference between γ_1 and γ_2 is small, relatively large values for h_2 can be obtained for small pressure differences, $p_A - p_B$.

值得注意的是，压差只与 h_2 有关，与 h_1 无关。如果 γ_1 和 γ_2 相差较小，即使测量较小的压差，也可以得到较大 h_2 的值。

(b) The specific value of the pressure difference for the data given is

(b) 对于给定的数据，压差的具体值为

$$p_A - p_B = 0.6 \times (15.60 - 8.60) = 4.20 \text{ (kPa)}$$

2.7.3 Inclined-Tube Manometer

2.7.3 倾斜管压力计

In order to measure small pressure changes, a manometer shown in Fig. 2.21 is often used. One leg of the manometer is inclined at an angle θ, and the differential reading along the inclined tube is l_2. The pressure difference $p_A - p_B$ can be expressed as

为了测量微小的压力变化，通常经常采用图 2.21 所示的压力计。压力计的一根管子倾斜 θ 角，沿着倾斜管的读数差为 l_2，则压差 $p_A - p_B$ 可表示为

$$p_A + \gamma_1 h_1 - \gamma_2 l_2 \sin\theta - \gamma_3 h_3 = p_B$$
$$p_A - p_B = \gamma_2 l_2 \sin\theta + \gamma_3 h_3 - \gamma_1 h_1 \quad (2.18)$$

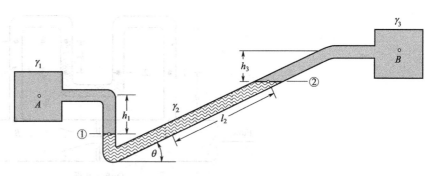

Fig. 2.21 Inclined-tube manometer
图 2.21 倾斜管压力计

It should be noted that the pressure difference between points ① and ② is caused by the vertical distance, $l_2 \sin\theta$, between the points. Therefore, even if the pressure difference is small, a large differential reading along the inclined tube can be obtained by using a small inclined angle. If pipes A and B contain gases, the pressure difference can be expressed as

值得注意的是点①和②之间的压差是由两点间的垂直距离 $l_2 \sin\theta$ 引起的。因此，即便压差很小，采用较小的倾斜角也可以得到较大的读数差。如果管 A 和 B 内为气体，则压差可表示为

$$p_A - p_B = \gamma_2 l_2 \sin\theta$$

$$l_2 = \frac{p_A - p_B}{\gamma_2 \sin\theta} \qquad (2.19)$$

where the contributions of the gas columns have been neglected. Equation (2.19) shows that, for a given pressure difference, the differential reading l_2 of the inclined-tube manometer is $1/\sin\theta$ times as large as that obtained by a conventional U-tube manometer.

Exercises

2.1 The water level in an open standpipe is 24m above the ground. What is the static pressure at a fire hydrant that is connected to the standpipe and located at ground level?

2.2 Blood pressure is commonly measured with a mercury manometer. A typical value for a human would be 120mmHg (systolic pressure)/70mmHg (diastolic pressure). What would these pressures be in pascals?

2.3 How much pressure will a diver be subjected to at a depth of 40 m in seawater?

2.4 A 0.4-m-diameter pipe is connected to a 0.02-m-diameter pipe and both are fixed in a position. Both pipes are horizontal with pistons at each end, and the space between the pistons is filled with water. If a force applied to the small piston is 100N, How much force will have to be applied to the large piston to balance the small position? (The friction can be neglected.)

2.5 A U-tube manometer is connected to a closed tank containing air and water as shown in Fig. 2.22. At the closed end of the manometer the air pressure is 110kPa. If the weight of the air columns in the manometer and tank can be neglected, determine the reading on the pressure gage.

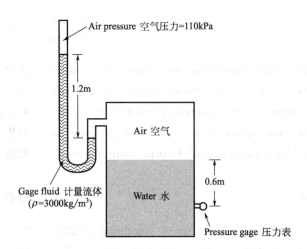

Fig. 2.22　Figure for the exercise 2.5
图 2.22　习题 2.5 图

2.6　For the stationary fluid shown in Fig. 2.23, the pressure at point B is 40kPa greater than that at point A. Determine the density of the gage fluid.

2.6　对于如图 2.23 所示的静止液体，点 B 的压力比点 A 大 40kPa。试确定计量流体的密度。

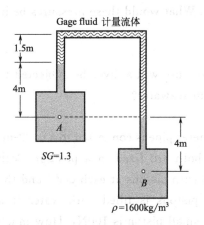

Fig. 2.23　Figure for the exercise 2.6
图 2.23　习题 2.6 图

2.7　For the inclined-tube manometer shown in Fig. 2.24, the pressure at point A is 40kPa. The fluid in both tanks is water, and the gage fluid in the manometer has a specific gravity of 2.4. What is the pressure at point B according to the values shown in this figure?

2.7　倾斜管压力计如图 2.24 所示，点 A 的压力为 40kPa。两箱体内的液体为水，压力计中的液体的比重为 2.4。根据图中所示的数值，点 B 的压力为多少？

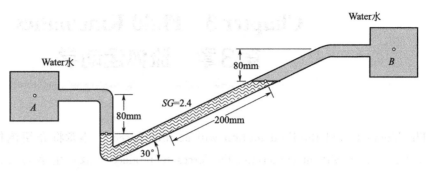

Fig. 2.24　Figure for the exercise 2.7
图 2.24　习题 2.7 图

2.8　An open tank has a vertical rectangular gate that is 5m high and 2m wide, and hinged at one end, as shown in Fig. 2.25. On one side it contains gasoline with a density $\rho = 800 \text{kg/m}^3$ at a depth of 5m. Water is slowly added to another side of the tank. At what depth, h, will the gate start to open?

2.8　一敞口储罐中间有一高 5m 宽 2m 的闸门，底端用铰链连接，如图 2.25 所示。一侧装有深 5m，密度为 800kg/m³ 的汽油，向另一侧缓慢加水。当水达到多深时，闸门将开启？

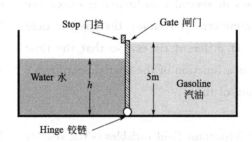

Fig. 2.25　Figure for the exercise 2.8
图 2.25　习题 2.8 图

2.9　A wooden cube 1m×1m×1m ($\rho = 780 \text{kg/m}^3$) floats in a tank of water. How much of the cube extends above the water surface? If the air pressure at the water surface increases to 8kPa, how much of the cube will extend above the water surface? Explain the reason for your answer.

2.9　一个密度为 780kg/m³、大小为 1m×1m×1m 的木块漂浮在水槽中，露在水面上的部分有多大？如果水面上大气压加压到 8kPa，此时露在水面上的部分多大？请解释原因。

Chapter 3　Fluid Kinematics
第 3 章　流体运动学

The kinematics of the fluid motion will be introduced in this chapter without concerning the forces to produce the motion. The analysis of the forces necessary to produce the motion (the dynamics of the fluid motion) will be discussed in detail in chapter 4.

本章将介绍流体运动学，而不涉及产生运动的力。产生运动所需的力的分析（流体的动力学）将在第 4 章中详细讨论。

3.1　The Velocity Field

3.1　速度场

At a given time point, any fluid property (such as density, pressure, velocity and acceleration) can be expressed as a function of the location of fluid. The function expresses in special coordinates is called the field of flow parameters. Of course, the specific field may be different at different times, so that the flow field is not only as a function of the spatial coordinates but also a function of time.

某一时刻，任何流体的属性（如密度、压力、速度和加速度）可以表示为流体位置的函数，这种用空间坐标表示的函数称为流动参量的场。当然，不同时刻的场可能不同，所以流场不仅是空间坐标的函数，还是时间的函数。

One of the most important fluid variables is the velocity field, which can be expressed as

速度场是最重要的流体变量之一，它可表示为

$$\mathbf{v} = v_x(x,y,z,t)\mathbf{i} + v_y(x,y,z,t)\mathbf{j} + v_z(x,y,z,t)\mathbf{k}$$

where v_x, v_y and v_z are the components of the velocity vector along the x, y and z directions, respectively. By definition, the velocity of a particle is the time rate of the position vector change for the particle. As illustrated in Fig. 3.1, the position vector, $\mathbf{r}_A$, of particle A is a function of time. The time derivative of this position gives the velocity of the particle, $d\mathbf{r}_A/dt = \mathbf{v}_A$. By writing the velocity for all particles, the field representation of the velocity vector $\mathbf{v} = \mathbf{v}(x, y, z, t)$ can be obtained.

式中，v_x、v_y 和 v_z 是速度矢量在 x、y 和 z 三个方向上的分量。根据定义，流体质点的速度是位置矢量的时间变化率。如图 3.1 所示，质点 A 的位置矢量为 $\mathbf{r}_A$ 是时间的函数。位置矢量对时间的导数即为质点的速度，$d\mathbf{r}_A/dt = \mathbf{v}_A$。写出所有质点速度即可得到速度矢量的流场描述，$\mathbf{v} = \mathbf{v}(x, y, z, t)$。

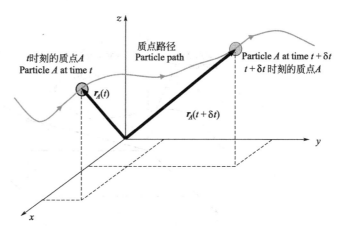

Fig. 3.1 Position vector of the particle
图 3.1 质点的位置矢量

The velocity **v** has both a direction and a magnitude, and its magnitude is

速度 **v** 具有方向和大小，其大小为

$$v=|\mathbf{v}|=(v_x^2+v_y^2+v_z^2)^{1/2}.$$

3.1.1 Eulerian Method and Lagrangian Method

3.1.1 欧拉法和拉格朗日法

There are two general approaches in analyzing fluid mechanics problems. The first one is called the Eulerian method. In this case, the fluid properties (pressure, density, velocity, etc.) are described as functions of space and time. From this method the flow information can be obtained in terms of what happens at fixed points in space as fluid flows past those points. The second method is called the Lagrangian method. In this case, fluid properties associate with individual fluid particles as a function of time are studied as they move.

分析流体力学问题有两种常用的方法。第一种方法称为欧拉法，这种方法将流体的属性（压力、密度和速度等）描述为空间和时间的函数，采用这种方法可以得到空间点上的流体流动信息。第二种方法称为拉格朗日法，该方法跟踪流体质点运动，研究流体质点属性随时间的变化。

The difference between the two methods can be illustrated in the example of temperature field, as shown in Fig. 3.2. In the Eulerian method a temperature measuring device may be fixed at any location (point 0: $x=x_0$, $y=y_0$, $z=z_0$) and the temperature will be recorded at different times. Then the temperature, T, for that location can be described as $T=T(x_0, y_0, z_0, t)$. If there are numerous points for measuring temperature, the temperature field can be expressed as $T=T(x, y, z, t)$.

如图 3.2 所示温度场的例子可以说明这两种方法的区别。如果采用欧拉方法，在某一位置（0 点：$x=x_0$, $y=y_0$, $z=z_0$）安装一个测温装置，记录不同时间的温度，则该位置的温度可描述为 $T=T(x_0, y_0, z_0, t)$。如果有许多测温点，则温度场可表示为 $T=T(x, y, z, t)$。

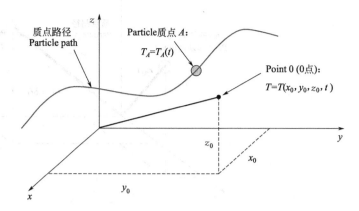

Fig. 3.2 Eulerian and Lagrangian descriptions of temperature field
图 3.2 温度场的欧拉和拉格朗日描述

In the Lagrangian method, the temperature measuring device may follow a particular fluid particle (particle A) and record that particle's temperature. As a result, the moving particle's temperature is obtained as a function of time, $T_A = T_A(t)$. If there are enough such measuring devices, the temperature of all fluid particles can be determined. If enough information in Eulerian form is available, Lagrangian information can be derived from the Eulerian data, and vice versa.

In fluid mechanics, the Eulerian method is usually used to describe a flow. But for some specific problems, the Lagrangian method is more convenient. For example, in some numerical simulations, the motion of individual fluid particles are determined by the interactions among the particles, thereby the motion is described using Lagrangian method.

若采用拉格朗日法，则在某一特定的流体质点 A 上安装一个测温装置，随质点运动并记录温度，该质点的温度可以描述为 $T_A = T_A(t)$。如果有足够的测温装置，可以确定所有流体质点温度。欧拉法和拉格朗日法之间可以相互转换。

流体力学中通常采用欧拉法描述流体流动。然而，对于某些特定的问题，采用拉格朗日法更为方便。例如，在一些数值模拟中，单个流体质点的运动由质点间的相互作用决定，因此用拉格朗日法描述。

3.1.2 One-, Two- and Three-Dimensional Flows

Generally, a fluid flow is a rather complex three-dimensional, time-dependent phenomenon. However, in many situations it is reasonable to simplify the problem based on some assumptions without reducing the needed accuracy. One of these simplifications is to approximate a three-dimensional flow as a one- or two-dimensional flow.

3.1.2 一维、二维和三维流动

通常，流体流动是相当复杂的三维时变现象。然而，在许多情况下，基于一些假设可将问题进行合理的简化，不降低所需的精度。简化方法之一就是将三维流动简化为一维或二维流动。

If one of the velocity components is small compared with the other two, it may be reasonable to neglect the smaller component and then the velocity is simplified to be $\mathbf{v}=v_x\mathbf{i}+v_y\mathbf{j}$.

Sometimes it is possible to further simplify a flow as a one-dimensional flow field, i.e., $\mathbf{v}=v_x\mathbf{i}$. Although very few flows are one-dimensional, it is still reasonable to treat some flow fields as the one-dimensional flow.

3.1.3 Steady and Unsteady Flows

In reality, flow parameters of almost all flows will vary with time thus they are unsteady flows. Obviously unsteady flows are usually more difficult to be analyzed than steady flows, and the unsteady flows include nonperiodic flow, periodic flow and random flow.

The flow of water through a faucet being turning off is just a nonperiodic unsteady flow. In this relatively slow process, it is not necessary to consider the forces generated by the unsteady effects. But if the water is turned off suddenly, the unsteady effects can not be ignored, for example the water hammer effect due to the sudden blocking-up will damage the pipe or the device connected.

In some flows the unsteady effects may be periodic. The periodic pressure fluctuation in the centrifugal pump is such an example. The unsteady effects repeat regularly at the frequency of the rotation of impeller.

In many situations the unsteady character of a flow is quite random. That is, there is no repeatable sequence or regular variation to the unsteady character. This behavior often occurs in turbulent flow.

It should be noted that to characterize a steady or unsteady flow is always based on a fixed space. For steady flow, all fluid properties, such as velocity, temperature, density, etc., at any fixed point are independent of time. But in terms of a fluid particle those properties may change with time as it flows about, even in steady flow. For example, the temperature of the steady exhaust gas flow may be constant at the chimney outlet, but the temperature of a fluid particle that left the chimney one minute ago is lower now than it was when it left the chimney, even though the flow is steady.

3.1.4 Streamlines, Streaklines and Pathlines

Although fluid motion can be quite complicated, there are various concepts that can be used to help in the analysis and visualization of flow fields. To this end streamlines, streaklines and pathlines are used in flow analysis. The streamline is often used in flow field analysis while the streakline and pathline are often used in experiments.

A streamline is a line that is everywhere tangent to the velocity. For steady flow, the streamlines are fixed lines in space. For unsteady flow, the streamlines may change shape with time. Streamlines can be obtained by integrating the tangential equations of the velocity. For two-dimensional flows, the slope of the streamline, dy/dx, is equal to the tangent value of the angle that the velocity vector makes with the x axis.

$$\frac{dx}{dy} = \frac{v_x}{v_y} \quad (3.1)$$

If the velocity field function is related to x and y, the equation of the streamlines can be obtained by integrating Eq. (3.1).

For unsteady flow, streamlines are very difficult to be produced experimentally. The observation of tracer injected into a flow can provide useful information, but it is not necessarily information about the streamlines.

位置而言的。对于定常流动，任意固定位置的所有流体属性（如速度、温度、密度等）均与时间无关。然而对于特定的流体质点，即使在定常流动中，流体属性仍可能随着质点的流动而改变。例如，烟囱出口的排出温度是常数，即使流动是稳定的，一分钟前排出的流体质点的温度要低于它刚排出时的温度。

3.1.4 流线、纹线和迹线

尽管流体运动非常复杂，但有许多概念可以帮助流场的分析和可视化。为此，将流线、纹线和迹线应用于流动分析中。流线主要用于流场分析，而纹线和迹线主要用于实验。

流线是一条处处与速度相切的线。对于定常流动，流线是空间固定的线。对于非定常流动，流线的形状可能会随时间变化。对速度切线方程进行积分可以得到流线。对于二维流动，流线的斜率 dy/dx 等于速度与 x 轴夹角的正切值。

如果速度场函数与 x 和 y 有关，对式（3.1）进行积分可以得到流线的方程。

对于非定常流动，难以通过实验得到流线。通过观察示踪剂可以得到一些有用的信息，但这未必就是关于流线的信息。

【EXAMPLE 3.1】 Determine the streamlines for the two-dimensional steady flow $\mathbf{v}=a(x\mathbf{i}+y\mathbf{j})$.

SOLUTION Since $v_x = ax$ and $v_y = ay$, streamlines are obtained from the equation

$$\frac{dy}{dx} = \frac{v_y}{v_x} = \frac{ay}{ax} = \frac{y}{x}$$

By separating the variables and integrating

$$\int \frac{dy}{y} = \int \frac{dx}{x} \quad \text{or} \quad \ln y = \ln x + C \quad \text{or} \quad \frac{y}{x} = C$$

where C is a constant. Different value of C correspond to different streamlines in the x-y plane. Thus, the equation for the streamlines of this flow is $\psi = \frac{y}{x}$. The streamlines in the first quadrant of the x-y plane are shown in Fig. 3.3.

【例题 3.1】 试确定二维定常流动 $\mathbf{v}=a(x\mathbf{i}+y\mathbf{j})$ 的流线。

解 由于 $v_x = ax$，$v_y = ay$，通过求解如下方程可得到流线

分离变量并积分得

式中，C 为常数。取不同的 C 值，可以在 x-y 平面上得到不同的流线，因此，流线方程为 $\psi = \frac{y}{x}$。x-y 平面第一象限的流线如图 3.3 所示。

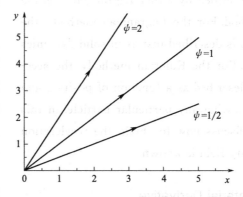

Fig. 3.3 Figure for the example 3.1
图 3.3 例题 3.1 图

A streakline consists of all particles in a flow that has previously passed through a common point. It can be obtained by taking photographs of marked particles that passed through a given point in the flow field. In experiment a streakline can be produced by continuously injecting marked fluid (such as smoke in air or dye in water) at a given point. For steady flow, the streakline coincides with the streamline through the injection point. For unsteady flow, particles injected at the same point at different times do not necessarily move

纹线由流经同一点的所有质点组成，它可以通过拍摄经过流场中给定位置的质点的照片得到。实验中，在给定位置连续喷射标记流体（如：在空气中喷射烟雾，或在水中喷射染料）可以产生纹线。对于定常流动，纹线和经过喷射点的流线重合。对于非定常流动，同一位置、不同时刻注入的质点不一定沿相同的路线运动，

along the same path. An instantaneous photograph of the marked fluid indicates the streakline at that instant, but it would not necessarily coincide with the streamline.

A pathline is the path of a given particle as it flows from one point to another. The pathline can be obtained by taking a time exposure photograph of a marked fluid particle.

For steady flow, pathlines, streamlines and streaklines coincide with each other. For unsteady flow, none of these three types of lines need to be the same.

3.2　The Acceleration Field

As mentioned in the previous section, the fluid motion can be described by either Lagrangian method or Eulerian method. For the Lagrangian method, the fluid acceleration is described just as in solid dynamics for each particle. For the Eulerian method, the acceleration field is described as a function of position and time without following any particular particle. In this section we will discuss how to obtain the acceleration field if the velocity field is known.

3.2.1　The Material Derivative

Figure 3.4 shows a fluid particle moving along its pathline, and its velocity, $\mathbf{v}_p$ for particle p, is a function of its location and the time. That is

$$\mathbf{v}_p = \mathbf{v}_p(\mathbf{r}_p, t) = \mathbf{v}_p(x_p, y_p, z_p, t)$$

where $x_p = x_p(t)$, $y_p = y_p(t)$ and $z_p = z_p(t)$. By definition, the acceleration of a particle is the time rate of change of its velocity. Since the velocity of a particle may change with time and its position, the acceleration of particle p, $\mathbf{a}_p$, can be described as

3.2　加速度场

如前所述，流体的运动可以采用拉格朗日法或者欧拉法进行描述。对于拉格朗日法，每一个质点加速度的描述就像固体动力学中一样。对于欧拉法，把加速度场描述为位置和时间的函数，而不跟踪任何特定的质点。本节我们将讨论如何由已知速度场获得加速度场。

3.2.1　物质导数

图 3.4 所示为一个流体质点沿它的迹线的运动，质点 p 的速度 $\mathbf{v}_p$ 为位置和时间的函数，即

式中，$x_p = x_p(t)$，$y_p = y_p(t)$，$z_p = z_p(t)$。根据定义，质点的加速度为速度的时间变化率，由于质点速度随时间和位置变化，因此，质点 p 的加速度 $\mathbf{a}_p$ 可表示为

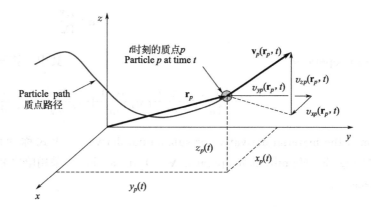

Fig. 3.4 Velocity and position of particle p at time t
图 3.4 t 时刻质点 p 的速度和位置

$$\mathbf{a}_p(t) = \frac{d\mathbf{v}_p}{dt} = \frac{\partial \mathbf{v}_p}{\partial t} + \frac{\partial \mathbf{v}_p}{\partial x}\frac{\partial x_p}{\partial t} + \frac{\partial \mathbf{v}_p}{\partial y}\frac{\partial y_p}{\partial t} + \frac{\partial \mathbf{v}_p}{\partial z}\frac{\partial z_p}{\partial t} \quad (3.2)$$

Since the velocity components can be expressed as $v_{xp} = \partial x_p/\partial t$, $v_{yp} = \partial y_p/\partial t$ and $v_{zp} = \partial z_p/\partial t$, Eq. (3.2) becomes

由于速度的分量可表示为 $v_{xp} = \partial x_p/\partial t$, $v_{yp} = \partial y_p/\partial t$, $v_{zp} = \partial z_p/\partial t$, 则式(3.2)变为

$$\mathbf{a}_p = \frac{\partial \mathbf{v}_p}{\partial t} + v_{xp}\frac{\partial \mathbf{v}_p}{\partial x} + v_{yp}\frac{\partial \mathbf{v}_p}{\partial y} + v_{zp}\frac{\partial \mathbf{v}_p}{\partial z}$$

Since the above equation is valid for any particle, the acceleration field can be written as

上式对任何质点均成立，所以加速度场可以表示为

$$\mathbf{a} = \frac{\partial \mathbf{v}}{\partial t} + v_x\frac{\partial \mathbf{v}}{\partial x} + v_y\frac{\partial \mathbf{v}}{\partial y} + v_z\frac{\partial \mathbf{v}}{\partial z} \quad (3.3)$$

Equation (3.3) can be written in scalar form as

式(3.3) 写成标量的形式为

$$\left.\begin{array}{l} a_x = \dfrac{\partial v_x}{\partial t} + v_x\dfrac{\partial v_x}{\partial x} + v_y\dfrac{\partial v_x}{\partial y} + v_z\dfrac{\partial v_x}{\partial z} \\[6pt] a_y = \dfrac{\partial v_y}{\partial t} + v_x\dfrac{\partial v_y}{\partial x} + v_y\dfrac{\partial v_y}{\partial y} + v_z\dfrac{\partial v_y}{\partial z} \\[6pt] a_z = \dfrac{\partial v_z}{\partial t} + v_x\dfrac{\partial v_z}{\partial x} + v_y\dfrac{\partial v_z}{\partial y} + v_z\dfrac{\partial v_z}{\partial z} \end{array}\right\} \quad (3.4)$$

where a_x, a_y and a_z are the components of the acceleration in x, y and z directions, respectively.

式中, a_x、a_y 和 a_z 分别为加速度在 x、y 和 z 三个方向上的分量。

For simplicity the above equations can be written in shorthand notation as

为简便起见，上式可以简记为

$$\mathbf{a} = \frac{D\mathbf{v}}{Dt}$$

where the operator

其中，算子为

$$\frac{D(\ )}{Dt} = \frac{\partial(\ )}{\partial t} + v_x \frac{\partial(\ )}{\partial x} + v_y \frac{\partial(\ )}{\partial y} + v_z \frac{\partial(\ )}{\partial z} \quad (3.5)$$

is termed the material derivative or substantial derivative. Using the Hamiltonian operator $\nabla(\ )$ it can be shortened as

上式称为随体导数或物质导数，采用哈密尔顿算子，可简记为

$$\frac{D(\ )}{Dt} = \frac{\partial(\ )}{\partial t} + \mathbf{v} \cdot \nabla(\ ) \quad (3.6)$$

3.2.2 Unsteady Effects

3.2.2 非定常作用

As can be seen from Eq. (3.5), the material derivative formula is composed of the term of time derivative, $\partial(\ )/\partial t$, and the term of spatial derivatives, $\partial(\ )/\partial x$, $\partial(\ )/\partial y$ and $\partial(\ )/\partial z$. The time derivative is called the local derivative, which represents unsteady effects of the flow. If the parameter involved is the acceleration, $\partial \mathbf{v}/\partial t$ is termed the local acceleration. For steady flow, the time derivative is zero, that is, $\partial(\ )/\partial t = 0$, and there is no change in flow parameters at a fixed point in space. However, for a fluid particle those parameters may change as it moves about.

由式（3.5）可见，物质导数公式由时间导数 $\partial(\ )/\partial t$ 和空间导数 $\partial(\ )/\partial x$、$\partial(\ )/\partial y$ 和 $\partial(\ )/\partial z$ 组成。时间导数称作局部导数，它表示流动的非定常作用。如果所涉及的参数是加速度，则 $\partial \mathbf{v}/\partial t$ 称为局部加速度。对于定常流动，时间导数为零，即 $\partial(\ )/\partial t = 0$，空间固定点上所有流动参数都不变化。然而，对于流体质点，这些参数可能会随其运动而变化。

For an unsteady flow, flow parameters at any location may change with time. For example, an unstirred ($\mathbf{v}=0$) cup of hot water will cool down due to heat transfer to its surrounding. That is, $DT/Dt = \partial T/\partial t + \mathbf{v} \cdot \nabla T = \partial T/\partial t < 0$. Similarly, a fluid particle may have acceleration as a result of the unsteady effect of the flow. As shown in Fig. 3.5, the flow is assumed to be spatially uniform throughout the constant diameter pipe. That is, $\mathbf{v} = v_0(t)\mathbf{i}$ at all points in the pipe. The value of the acceleration depends on whether v_0 is being increased ($\partial v_0/\partial t > 0$) or decreased ($\partial v_0/\partial t < 0$). Unless v_0 is independent of time (v_0 = constant) there will be

对于非定常流动，任何位置的流动参数都可能随时间变化。例如，一杯静止的热水（$\mathbf{v}=0$）因向周围传热冷却下来，即 $DT/Dt = \partial T/\partial t + \mathbf{v} \cdot \nabla T = \partial T/\partial t < 0$。类似地，由于流动的非定常作用，流体质点会有加速度。如图3.5所示，等直径管内流动空间均匀分布，即在管内各点 $\mathbf{v} = v_0(t)\mathbf{i}$。加速度的值取决于 v_0 增加（$\partial v_0/\partial t > 0$）还是减小（$\partial v_0/\partial t < 0$）。除非 v_0 与时间无关（v_0 = 常数），加速度都会

an acceleration. Thus, the acceleration field, $\mathbf{a}=(\partial v_0/\partial t)\mathbf{i}$, is uniform throughout the entire flow, although it may vary with time. Since $\partial v_x/\partial x=0$ and $v_y=v_z=0$, there is no acceleration due to the spatial variations of velocity for this flow, that is

存在。因此，整个流动的加速度场是均匀的，尽管它可能随时间变化。由于 $\partial v_x/\partial x=0$ 并且 $v_y=v_z=0$，对于该流动并不存在因速度的空间变化而产生的加速度，即

$$\mathbf{a}=\frac{\partial \mathbf{v}}{\partial t}+v_x\frac{\partial \mathbf{v}}{\partial x}+v_y\frac{\partial \mathbf{v}}{\partial y}+v_z\frac{\partial \mathbf{v}}{\partial z}=\frac{\partial \mathbf{v}}{\partial t}=\frac{\partial v_0}{\partial t}\mathbf{i}$$

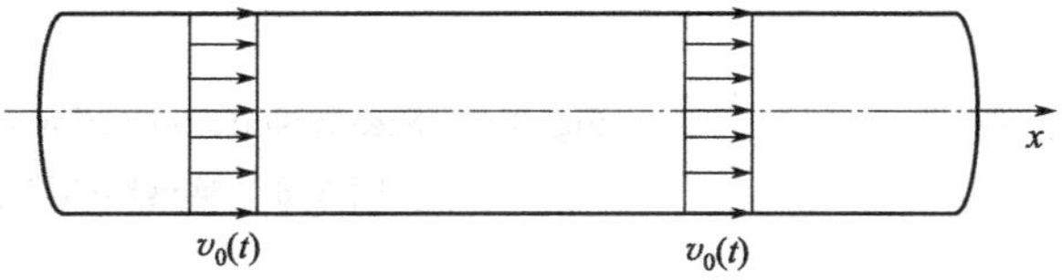

Fig. 3.5 Uniform, unsteady flow in a constant diameter pipe

图 3.5 定径管内均匀非定常流动

3.2.3 Convective Effects

3.2.3 对流作用

The spatial derivative in Eq. (3.5), known as the convective derivative, represents that the variation of a flow property of a fluid particle initiated by its motion. Since there is a gradient $\mathbf{\nabla}(\ )$ in Eq. (3.6), the spatial rate of change of a parameter will exist no matter the flow is steady or unsteady. The acceleration given by the term $(\mathbf{v}\cdot\mathbf{\nabla})\mathbf{v}$ is termed the convective acceleration.

式(3.5)中的空间导数称作对流导数，它表示质点的流动特性因质点运动而变化。由于式(3.6)中存在梯度 $\mathbf{\nabla}(\ )$，不论流动是否定常，参数的空间变化率都会存在。由 $(\mathbf{v}\cdot\mathbf{\nabla})\mathbf{v}$ 给出的这部分加速度称作对流加速度。

As is illustrated in Fig. 3.6, the water entering the heat exchanger is always the same cold temperature and the water leaving the heater is always the same hot temperature. The flow is steady. However, the temperature, T, of each water particle increases as it passes through the heat exchanger, and as a result, $T_{out}>T_{in}$. Thus, $DT/Dt\neq 0$ because of the convective term in the total derivative of the temperature. That is, $\partial T/\partial t=0$, but $v_x(\partial T/\partial x)\neq 0$ (where x is directed along the streamline), since the temperature gradient along the streamline is not zero. Therefore, for this instance $DT/Dt=v_x(\partial T/\partial x)$ even though the flow is steady.

如图 3.6 所示的换热器，进入换热器的冷水温度不变，流出换热器的热水温度也不变，流动是定常的。然而，随着水通过换热器，温度不断升高，$T_{out}>T_{in}$。因此，由于温度导数中的对流项存在，$DT/Dt\neq 0$。也就是说，虽然 $\partial T/\partial t=0$，但是由于沿流线方向温度梯度不为零，$v_x(\partial T/\partial x)\neq 0$（$x$ 为流线方向）。因此，对于这种情况，即使流动定常，$DT/Dt=v_x(\partial T/\partial x)$。

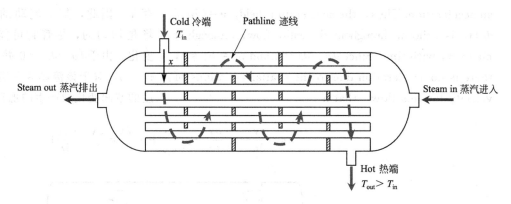

Fig. 3.6 Steady-state operation of a heat exchanger
图 3.6 换热器的稳态运行

There also exists fluid acceleration in the flow in a pipe with variable diameter, as shown in Fig. 3.7. The flow is assumed to be steady and one-dimensional, and the velocity increases from v_1 to v_2 as the fluid flows from x_1 to x_2. Therefore, even though $\partial \mathbf{v}/\partial t = 0$, there still exists acceleration $a_x = v_x(\partial v_x/\partial x)$. In the range of $x_1 < x < x_2$, $\partial v_x/\partial x > 0$ so that $a_x > 0$, the fluid accelerates. In the range of $x_2 < x < x_3$, $\partial v_x/\partial x < 0$ so that $a_x < 0$, the fluid decelerates. If $v_1 = v_3$, the amount of acceleration precisely offsets the amount of deceleration.

如图 3.7 所示，变径管内流动也存在加速度。假设流动为一维稳定流动，从 x_1 到 x_2，速度从 v_1 增加到 v_2，因此，即使 $\partial \mathbf{v}/\partial t = 0$，仍存在加速度 $a_x = v_x(\partial v_x/\partial x)$。在 $x_1 < x < x_2$ 范围内，$\partial v_x/\partial x > 0$，所以 $a_x > 0$，流体加速；在 $x_2 < x < x_3$ 范围内，$\partial v_x/\partial x < 0$，所以 $a_x < 0$，流体减速。若 $v_1 = v_3$，加速量正好抵消减速量。

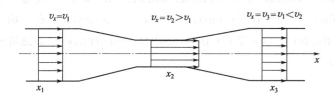

Fig. 3.7 Uniform steady flow in a variable area pipe
图 3.7 变截面管道内的均匀定常流动

【EXAMPLE 3.2】 Determine the acceleration field for the steady and two-dimensional flow field discussed in Example 3.1.

【例题 3.2】 试确定例 3.1 中的定常二维流动的加速度场。

SOLUTION The acceleration is generally expressed as

解 加速度一般表达为

$$\mathbf{a} = \frac{D\mathbf{v}}{Dt} = \frac{\partial \mathbf{v}}{\partial t} + (\mathbf{v} \cdot \nabla)\mathbf{v} = \frac{\partial \mathbf{v}}{\partial t} + v_x \frac{\partial \mathbf{v}}{\partial x} + v_y \frac{\partial \mathbf{v}}{\partial y} + v_z \frac{\partial \mathbf{v}}{\partial z} \quad (3.7)$$

where the velocity is given by $\mathbf{v} = a(x\mathbf{i} + y\mathbf{j})$, so that $v_x = ax$ and $v_y = ay$. For steady flow, $\partial()/\partial t = 0$. For

式中，速度 $\mathbf{v} = a(x\mathbf{i} + y\mathbf{j})$，所以 $v_x = ax$，$v_y = ay$。对于定常流动，

two-dimensional flow, $v_z = 0$ and $\partial()/\partial z = 0$. Therefore, Eq. (3.7) becomes

$$\mathbf{a} = v_x \frac{\partial \mathbf{v}}{\partial x} + v_y \frac{\partial \mathbf{v}}{\partial y} = \left(v_x \frac{\partial v_x}{\partial x} + v_y \frac{\partial v_x}{\partial y} \right) \mathbf{i} + \left(v_x \frac{\partial v_y}{\partial x} + v_y \frac{\partial v_y}{\partial y} \right) \mathbf{j}$$

Hence, for this flow the acceleration is given by

$$\mathbf{a} = [(ax)a + (ay)0]\mathbf{i} + [(ax)0 + (ay)a]\mathbf{j}$$

or

$$a_x = a^2 x, \quad a_y = a^2 y$$

$\partial()/\partial t = 0$。对于二维流动，$v_z = 0$，$\partial()/\partial z = 0$。因此，式(3.7)变为

因此，对于该流动，加速度为

或

3.3 Fluid Element Kinematics

The mathematical description for the motion of fluid elements in a flow field will be discussed in this section. Assume there is a cubic fluid element moving from one position to another in a short time interval, as illustrated in Fig. 3.8. In the complex velocity field, the element not only translates from one position to another position but also experience volume change (linear deformation), rotation and shape change (angular deformation). Although these movements and deformations occur simultaneously, each of them can be considered separately.

3.3 流体微团运动

本节讨论流场中流体微团运动的数学描述。如图3.8所示，假设一个方形流体微团在很短的时间间隔内从一个位置运动到另一个位置。在复杂的速度场中，流体微团不仅发生了平移，还发生了体积变化（线变形）、旋转、形状变化（角变形）。尽管这些运动和变形是同时发生的，但可以分别进行考虑。

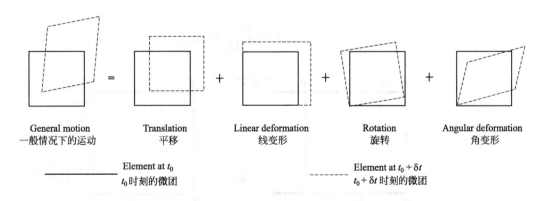

Fig. 3.8 Types of motion and deformation for a fluid element
图 3.8 流体微团的运动和变形类型

3.3.1 Linear Motion and Deformation

Translation is the simplest motion form, as illustrated in Fig. 3.9. In a small time interval a particle in the element will move from point O to point O'. If all points in the element have the same velocity, the element will simply translate from one position to another.

3.3.1 线运动和变形

如图 3.9 所示,平移是最简单的运动形式。很短的时间内质点从点 O 运动到点 O',如果微团中所有点都有相同的速度,则微团从一个位置平移到另一个位置。

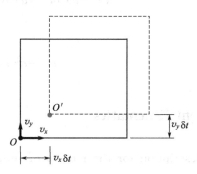

Fig. 3.9　Translation of a fluid element
图 3.9　流体微团的平移

However, because of the presence of velocity gradients, the element will generally be deformed and rotated as it moves. As shown in Fig. 3.10(a), if the x component of velocity of O and B is v_x, the x component of the velocity of nearby points A and C can be expressed as $v_x + (\partial v_x / \partial x) \delta x$. During the time interval δt, the element stretches by an amount $(\partial v_x / \partial x) \delta x \delta t$ as illustrated in Fig. 3.10(b). Thereby the volume change of the element ($\delta V = \delta x \delta y \delta z$) is

然而,由于速度梯度的存在,随着运动,微团将发生变形和旋转。如图 3.10(a) 所示,如果点 O 和点 B 在 x 方向的速度分量为 v_x,那么附近的点 A 和 C 在 x 方向的速度分量可以表达为 $v_x + (\partial v_x / \partial x) \delta x$。$\delta t$ 时间间隔内,微团伸长了 $(\partial v_x / \partial x) \delta x \delta t$,如图 3.10(b) 所示。因此,流体微团的体积变化量为

$$\delta V = \left(\frac{\partial v_x}{\partial x} \delta x \right) \delta y \delta z \delta t$$

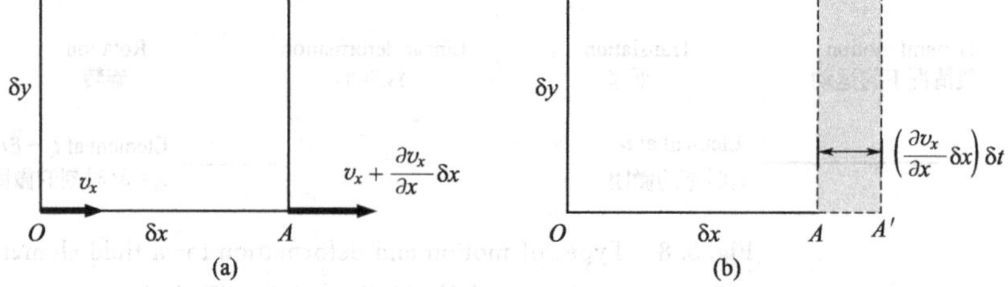

Fig. 3.10　Linear deformation of a fluid element
图 3.10　流体微团的线变形

and the rate of change of the volume per unit volume is   单位体积的体积变化率为

$$\frac{1}{\delta V}\frac{\mathrm{d}(\delta V)}{\mathrm{d}t} = \lim_{\delta t \to 0}\left[\frac{(\partial v_x/\partial x)\delta t}{\delta t}\right] = \frac{\partial v_x}{\partial x} \quad (3.8)$$

If other two velocity gradients $\partial v_y/\partial y$ and $\partial v_z/\partial z$ exist, we have   如果同时有$\partial v_y/\partial y$和$\partial v_z/\partial z$,则

$$\frac{1}{\delta V}\frac{\mathrm{d}(\delta V)}{\mathrm{d}t} = \frac{\partial v_x}{\partial x} + \frac{\partial v_y}{\partial y} + \frac{\partial v_z}{\partial z} = \nabla \cdot \mathbf{v} \quad (3.9)$$

This is the volumetric dilatation rate of fluid. Thus, it can be seen that the volume of a fluid may change as the element moves. However, for an incompressible fluid the volumetric dilatation rate is zero, since its density can not be changed. The derivatives $\partial v_x/\partial x$, $\partial v_y/\partial y$ and $\partial v_z/\partial z$ represent the velocity variation along its own direction, which only cause a linear deformation of the element.

上式即为流体的体积膨胀率。可见,随着微团的移动,流体的体积可能会发生变化。然而,对于不可压缩流体,由于流体的密度不变,体积膨胀率为零。速度在其自身方向上的变化($\partial v_x/\partial x$、$\partial v_y/\partial y$和$\partial v_z/\partial z$)只引起流体微团的线变形。

3.3.2 Angular Motion and Deformation

3.3.2 角运动和变形

For simplicity, only the motion in the x-y plane will be considered, but the results can be extended to three-dimensional cases. The rotation and angular deformation that caused by velocity variation are illustrated in Fig. 3.11(a). In a short time interval, δt, the line segments OA and OB will rotate to the new positions OA' and OB' by the angles of $\delta \alpha$ and $\delta \beta$, as shown in Fig. 3.11(b). The angular velocity of line segment OA, ω_{OA}, is

为简便起见,这里只考虑x-y平面内的运动,但结果可以推广到三维情况。速度变化引起的旋转和角变形如图3.11(a)所示,在δt时间内,线段OA和OB分别旋转角度$\delta \alpha$和$\delta \beta$到新的位置OA'和OB',如图3.11(b)所示。线段OA的角速度ω_{OA}为

$$\omega_{OA} = \lim_{\delta t \to 0}\frac{\delta \alpha}{\delta t}$$

For small angles   转角很小时

$$\tan\delta\alpha \approx \delta\alpha = \frac{(\partial v_y/\partial x)\delta x \delta t}{\delta x} = \frac{\partial v_y}{\partial x}\delta t \quad (3.10)$$

so that   因此

$$\omega_{OA} = \lim_{\delta t \to 0}\left[\frac{(\partial v_y/\partial x)\delta t}{\delta t}\right] = \frac{\partial v_y}{\partial x}$$

Note that if $\partial v_y/\partial x$ is positive, ω_{OA} will be counterclockwise. Similarly, the angular velocity of the line OB, ω_{OB}, is

注意：若$\partial v_y/\partial x$为正，则ω_{OA}为逆时针方向。类似地，OB的角速度ω_{OB}为

$$\omega_{OB} = \lim_{\delta t \to 0} \frac{\delta \beta}{\delta t}$$

For small angles

转角很小时

$$\tan\delta\beta \approx \delta\beta = \frac{(\partial v_x/\partial y)\delta y \delta t}{\delta y} = \frac{\partial v_x}{\partial y}\delta t \quad (3.11)$$

so that

因此

$$\omega_{OB} = \lim_{\delta t \to 0}\left[\frac{(\partial v_x/\partial y)\delta t}{\delta t}\right] = \frac{\partial v_x}{\partial y}$$

In this case if $\partial v_x/\partial y$ is positive, ω_{OB} will be clockwise. The rotational velocity, ω_z, of the element about the z axis is defined as the average of the angular velocities ω_{OA} and ω_{OB}. Thus, if counterclockwise rotation is regarded as the positive, it follows that

这种情况下，若$\partial v_x/\partial y$为正，则ω_{OB}为顺时针方向。微团绕z轴旋转的转速ω_z定义为角速度ω_{OA}和ω_{OB}的平均值。因此，如果认为逆时针旋转为正，可得

$$\omega_z = \frac{1}{2}\left(\frac{\partial v_y}{\partial x} - \frac{\partial v_x}{\partial y}\right) \quad (3.12)$$

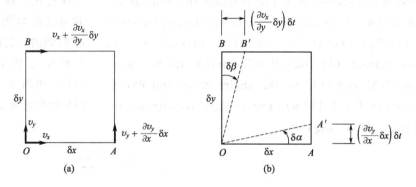

Fig. 3.11 Rotation and angular deformation of a fluid element
图 3.11 流体微团的旋转和角变形

Using the same analysis method, rotations of the fluid element about the x axis and the y axis can also be obtained as

采用同样的分析方法，得到流体微团绕x和y轴旋转的转速为

$$\omega_x = \frac{1}{2}\left(\frac{\partial v_z}{\partial y} - \frac{\partial v_y}{\partial z}\right) \quad (3.13a)$$

$$\omega_y = \frac{1}{2}\left(\frac{\partial v_x}{\partial z} - \frac{\partial v_z}{\partial x}\right) \quad (3.13b)$$

The angular velocity vector, **ω**, can be expressed in the form

角速度矢量 ω 可表示为

$$\boldsymbol{\omega} = \omega_x \mathbf{i} + \omega_y \mathbf{j} + \omega_z \mathbf{k} \quad (3.14)$$

It can be seen that **ω** is equal to one-half the convolution of the velocity vector. That is

可见，角速度 ω 等于速度矢量卷积的一半，即

$$\boldsymbol{\omega} = \frac{1}{2}\text{curl }\mathbf{v} = \frac{1}{2}\boldsymbol{\nabla} \times \mathbf{v} \quad (3.15)$$

where the vector operator $\boldsymbol{\nabla} \times \mathbf{v}$ is

$\boldsymbol{\nabla} \times \mathbf{v}$ 为

$$\boldsymbol{\nabla} \times \mathbf{v} = \begin{vmatrix} \mathbf{i} & \mathbf{j} & \mathbf{k} \\ \frac{\partial}{\partial x} & \frac{\partial}{\partial y} & \frac{\partial}{\partial z} \\ v_x & v_y & v_z \end{vmatrix} = \left(\frac{\partial v_z}{\partial y} - \frac{\partial v_y}{\partial z}\right)\mathbf{i} + \left(\frac{\partial v_x}{\partial z} - \frac{\partial v_z}{\partial x}\right)\mathbf{j} + \left(\frac{\partial v_y}{\partial x} - \frac{\partial v_x}{\partial y}\right)\mathbf{k}$$

The vorticity, **Ω**, is defined as a vector that is twice the angular velocity vector, that is

旋度 Ω 定义为两倍的角速度，即

$$\boldsymbol{\Omega} = 2\boldsymbol{\omega} = \boldsymbol{\nabla} \times \mathbf{v} \quad (3.16)$$

The vorticity is often used to describe the rotational characteristics of the fluid. If $\boldsymbol{\nabla} \times \mathbf{v} = 0$, the vorticity is zero, and such a flow field is called irrotational flow field.

旋度常用于描述流体的旋转特征。如果 $\boldsymbol{\nabla} \times \mathbf{v} = 0$，则旋度为零，这样的流场称为无旋流场。

3.4 System and Control Volume

3.4 系统和控制体

As with any matter, behaviors of fluids obey a set of fundamental physical laws, such as the conservation of mass, Newton's laws of motion and the laws of thermodynamics. For fluids, these governing laws can be applied on the system or the control volume. By definition, a system is a collection of matter of fixed identity. A control volume, on the other hand, is a volume in space through which fluid may flow.

和任何物质一样，流体的行为也服从基本物理定律，如质量守恒定律、牛顿运动定律和热力学定律。对于流体，这些定律可应用于系统或控制体上。根据定义，系统是指有确定性质的物质集合，而控制体是指流体流过的空间体。

A system always contains the unchangeable mass but has changeable size and shape. It may interact with its surroundings by means of the transfer of heat or the exertion of a force.

系统的质量不变，但尺寸和形状可变。它可以通过传热或力的作用与周围环境相互影响。

In the study of statics and dynamics, we usually identify a free-body firstly and then isolate it from its surroundings. Its surroundings are replaced by some equivalent actions and then the Newton's laws of motion can be applied. The free-body selected in fluid is just the system. Different from a solid that may deform but is still easy to be identified, fluids are quite difficult to be identified and traced because they can move quite freely. For example, it is easy to follow a piece of feather blown by wind, but difficult to follow a specific portion of air flowing in the sky.

Compared with the information given by a portion of fluid (a system), we are more interested in the forces on an object initiated by fluid flow. On this account the control volume approach has been used more frequently. Before performing the control volume analysis, a specific volume in space should be identified and then the fluid flow within, through or around that volume will be analyzed. In general, the control volume can be moved, but the control volumes concerned in this book are fixed and indeformable. The matter and the amount of mass within a control volume may change with time as fluids flow in or out. The control volume itself is just a specific geometry, independent of fluid flow.

Most fluid flow problems can be solved based on a fixed and nondeforming control volume. However, for some situations, it is better to use a moving or deforming control volume, as shown in Fig. 3.12.

在静力学和动力学研究中，首先需要确定一个自由体，然后将自由体与周围环境分离，并用等效作用代替它的周围环境，于是牛顿运动定律就可以使用了。所选取的这个自由体就是系统。固体虽然会变形，但依旧容易辨认。与固体不同，流体可自由流动，故很难辨认和跟踪。例如跟踪被风吹起的一片羽毛比较容易，但跟踪空中流动的一部分空气就很困难。

与部分流体（系统）的具体流动信息相比，我们更关心的是流体作用在物体上的力，因此，控制体方法使用的更多。进行控制体分析之前，先确定一个具体的流场空间，然后分析该空间内部、流经该空间以及该空间周围的流体流动。通常，控制体可以移动，但本书涉及的控制体是固定且不变形的。随着流体的流入或流出，控制体内的物质和质量可以随时间改变，但控制体本身是一个特定的几何体，与流动流体无关。

大多数流体问题可以采用固定的、不变形的控制体进行求解。但有些情况，使用移动的或形状变化的控制体更方便，如图3.12所示。

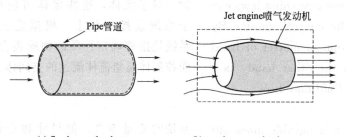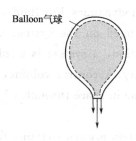

Fig. 3.12 Typical control volumes

图 3.12 典型的控制体

The relationship between a system and a control volume is, in a certain extent, similar to that between the Lagrangian and Eulerian flow descriptions. In the system or Lagrangian description, attention is paid to a portion of fluid with its behavior being investigated as it moves. In the control volume or Eulerian description, attention is paid to the flow behavior of fluid at a fixed position.

All of the laws governing the motion of fluid are stated in their basic form in terms of a system approach. For example, the mass of a system remains constant, or the time rate of change of momentum of a system is equal to the sum of all the forces acting on the system. To use the governing equations in a control volume approach, the laws in system approach must be rephrased in control volume approach. To do this we will use the Reynolds transport theorem.

3.5 Reynolds Transport Theorem

Sometimes it is necessary to use the laws in system concepts to govern fluid motion while other times the laws in control volume concepts are used. The Reynolds transport theorem provides a link between presentations in system concepts and control volume concepts.

Obviously, any physical law should be expressed by a set of physical parameters. Let B represent any of the fluid parameters and b represent the amount of that parameter per unit mass. That is

$$B = mb$$

where m is the mass of the fluid. For example, if $B = mv^2/2$ represents the kinetic energy of the mass, $b = v^2/2$ represents the kinetic energy per unit mass. The parameters B and b may be scalars or vectors. Thus, if $\mathbf{B} = m\mathbf{v}$ represents the momentum of the mass, $\mathbf{b} = \mathbf{v}$ represents the momentum per unit mass (the velocity).

3.5 雷诺输运定理

系统和控制体的关系一定程度上与拉格朗日法和欧拉法的流动描述关系相似。系统的观点和拉格朗日法着眼于跟随流体流动,研究流动行为。控制体的观点和欧拉法则着眼于研究某一固定位置上流体的流动行为。

控制流体运动的所有定律都是基于系统的方法表述的,如系统的质量不变,系统的动量变化率等于作用在系统上的所有力之和。为了使用基于控制体方法的控制方程,必须将基于系统的定律表达成适用于控制体的形式。为此,要用到雷诺输运定理。

有时需要用基于系统概念的定律来控制流体运动,而有时又需要用基于控制体概念的定律,雷诺输运定理提供了两者之间联系的描述。

显然,任何物理定律都要采用物理参数来表示。令 B 表示任意流体参数,b 表示单位质量该参数的量,则

式中,m 为流体的质量。例如:若 $B = mv^2/2$ 表示一定质量所具有的动能,那么 $b = v^2/2$ 表示单位质量所具有的动能。B 和 b 可以是标量也可以是矢量。因此,如果 $\mathbf{B} = m\mathbf{v}$ 表示一定质量所具有的动量,那么 $\mathbf{b} = \mathbf{v}$ 表示单位质量所具有的动量,即速度。

The parameter B is termed an extensive property and the parameter b is termed an intensive property. For an infinitesimal fluid element with the size δV and mass $\rho \delta V$, the overall amount of B of the system can be obtained from an integration in the system

参数 B 称作广延属性，参数 b 称作强度属性。对于体积为 δV、质量为 $\rho\delta V$ 的流体微元，系统所具有的 B 的总量可以通过对整个系统进行积分而得

$$B_{\text{sys}} = \lim_{\delta V \to 0} \sum_i b_i (\rho_i \delta V_i) = \int_{\text{sys}} \rho b \, dV$$

The time rate of change of an extensive property of a fluid system, such as the time rate of change of the momentum of a system, the time rate of change of mass of a system, can be expressed as

流体系统广延属性的时间变化率（如：系统动量的时间变化率、系统质量的时间变化率等）可以表达为

$$\frac{dB_{\text{sys}}}{dt} = \frac{d\left(\int_{\text{sys}} \rho b \, dV\right)}{dt} \quad (3.17)$$

The time rate of change of an extensive property within a control volume can be expressed as

基于控制体的广延属性的时间变化率，它可以表达为

$$\frac{dB_{\text{cv}}}{dt} = \frac{d\left(\int_{\text{cv}} \rho b \, dV\right)}{dt} \quad (3.18)$$

where the subscript cv means the control volume. Although Eq. (3.17) and Eq. (3.18) appear to be similar, their physical meanings are quite different.

式中，下标 cv 表示控制体。尽管式（3.17）和式（3.18）形式上类似，但物理意义不同。

【EXAMPLE 3.3】 Gas leaks from the gas tank shown in Fig. 3.13. Discuss the differences between dB_{sys}/dt and dB_{cv}/dt when B represents mass.

【例题 3.3】 气体从煤气罐中泄漏，如图 3.13 所示。如果 B 表示质量，试讨论 dB_{sys}/dt 和 dB_{cv}/dt 之间的差别。

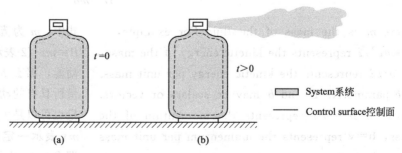

Fig. 3.13　Figure for the example 3.3
图 3.13　例题 3.3 图

SOLUTION Let B be the mass of the systerm m, then $b=1$. Equation (3.17) and Eq. (3.18) can be written as

$$\frac{dB_{sys}}{dt} \equiv \frac{dm_{sys}}{dt} = \frac{d\left(\int_{sys}\rho dV\right)}{dt}$$

$$\frac{dB_{cv}}{dt} \equiv \frac{dm_{cv}}{dt} = \frac{d\left(\int_{cv}\rho dV\right)}{dt}$$

The two equations represent the time rate of change of mass within the system and that within the control volume, respectively. In this case, the gas in the tank at $t=0$ is chosen as the system and the tank itself as the control volume. A short time later, part of the system has moved outside of the control volume as shown in Fig. 3.13(b), while the control volume does not move. The limit of integration is invariant for the control volume. However, the limit of integration for the system changes with time.

Obviously, since mass is conserved, the mass of the fluid in the system is constant, so that

$$\frac{d\left(\int_{sys}\rho dV\right)}{dt} = 0$$

However, in the control volume the amount of mass decreases continuously; that is

$$\frac{d\left(\int_{cv}\rho dV\right)}{dt} < 0$$

Clearly the meanings of dB_{sys}/dt and dB_{cv}/dt are different. For this example, $dB_{cv}/dt < dB_{sys}/dt$.

3.5.1 Derivation of the Reynolds Transport Theorem

The Reynolds transport theorem can be obtained by analyzing the flow through a fixed control volume as shown in Fig. 3.14. The outflow from the control volume from time t to $t+\delta t$ is denoted as volume Ⅱ, the

解 B 表示系统的质量 m,则 $b=1$,式(3.17)和式(3.18)可写为

以上两个方程分别表示系统内的质量变化率和控制体内的质量变化率。选择 $t=0$ 时刻煤气罐内的气体作为系统,煤气罐本身作为控制体。一段时间后,系统的一部分流出控制体,如图3.13(b)所示,而控制体没有移动。对于控制体,积分范围不变;而对于系统,积分范围是随时间变化的。

显然,由于质量是守恒的,系统内流体的质量为常数,故

然而,控制体内的质量不断减少,即

显然,dB_{sys}/dt 和 dB_{cv}/dt 的含义是不同的。本例中,$dB_{cv}/dt < dB_{sys}/dt$。

3.5.1 雷诺输运定理的推导

通过分析流经固定控制体的流体流动(如图3.14所示),可以得到雷诺输运定理。从 t 时刻到 $t+\delta t$ 时刻,流出控制体的流体表示为Ⅱ,

inflow is denoted as volume Ⅰ, and the control volume itself is denoted as cv. Thus, the system at time t consists of the fluid in section cv, that is "sys=cv", while at time $t+\delta t$ the system occupies section (cv−Ⅰ)+Ⅱ, that is "sys=cv−Ⅰ+Ⅱ". The control volume remains as section cv for all time.

流入控制体的流体表示为Ⅰ，控制体表示为cv。于是，t时刻系统由cv内的流体组成，即sys=cv。而在$t+\delta t$时刻，系统由(cv−Ⅰ)+Ⅱ内的流体组成，即sys=cv−Ⅰ+Ⅱ。控制体cv保持不变。

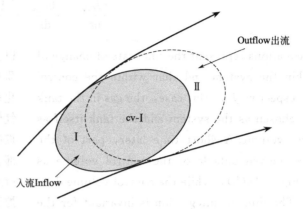

Fig. 3.14   Control volume and system
图 3.14   控制体和系统

If B is an extensive parameter of the system, then its value for the system at time t is

如果B是系统的广延参数，那么t时刻其值为

$$B_{sys}(t)=B_{cv}(t)$$

Its value at time $t+\delta t$ is

$t+\delta t$时刻系统的广延参数的值为

$$B_{sys}(t+\delta t)=B_{cv}(t+\delta t)-B_{Ⅰ}(t+\delta t)+B_{Ⅱ}(t+\delta t)$$

Thus, the change in the amount of B in the system in the time interval δt divided by this time interval is given by

δt时间间隔内系统广延参数B的变化量除以时间间隔可表达为

$$\frac{\delta B_{sys}}{\delta t}=\frac{B_{sys}(t+\delta t)-B_{sys}(t)}{\delta t}=\frac{B_{cv}(t+\delta t)-B_{Ⅰ}(t+\delta t)+B_{Ⅱ}(t+\delta t)-B_{sys}(t)}{\delta t}$$

At the initial time t, $B_{sys}(t)=B_{cv}(t)$. By substituting it into the above equation, the following expression can be obtained

在初始时刻t，$B_{sys}(t)=B_{cv}(t)$，将其代入上式变形后可得

$$\frac{\delta B_{sys}}{\delta t}=\frac{B_{cv}(t+\delta t)-B_{cv}(t)}{\delta t}-\frac{B_{Ⅰ}(t+\delta t)}{\delta t}+\frac{B_{Ⅱ}(t+\delta t)}{\delta t} \quad (3.19)$$

In the limit $\delta t \to 0$, the left side of Eq. (3.19) is equal to the time rate of change of B of a system and is denoted as DB_{sys}/Dt. This may represent the rate of change of mass, momentum, energy, or angular momentum of the system, depending on the choice of the parameter B.

当 $\delta t \to 0$ 时，式(3.19)的左边等于 B 的变化率，表示为 DB_{sys}/Dt。它可以表示系统的质量、动量、能量或角动量的变化率，具体取决于参数 B 的选择。

The first term on the right side of Eq. (3.19) represents the time rate of change of B within the control volume

式(3.19)右边的第一项为控制体内 B 的时间变化率

$$\lim_{\delta t \to 0} \frac{B_{cv}(t+\delta t) - B_{cv}(t)}{\delta t} = \frac{\partial B_{cv}}{\partial t} = \frac{\partial \left(\int_{cv} \rho b \, dV \right)}{\partial t} \quad (3.20)$$

The second term on the right side of Eq. (3.19) represents the rate of the extensive parameter B flows into the control volume across the control surface, and it is denoted as $\dot{B}_{in}$

式(3.19)右边第二项表示通过控制面流入控制体的广延参数 B 的速率，表示为 $\dot{B}_{in}$

$$\dot{B}_{in} = \lim_{\delta t \to 0} \frac{B_I(t+\delta t)}{\delta t} \quad (3.21)$$

The third term on the right side of Eq. (3.19) represents the rate of the extensive parameter B flows out of the control volume across the control surface, and it is denoted as $\dot{B}_{out}$

式(3.19)右边第三项表示通过控制面流出控制体的广延参数 B 的速率，表示为 $\dot{B}_{out}$

$$\dot{B}_{out} = \lim_{\delta t \to 0} \frac{B_{II}(t+\delta t)}{\delta t} \quad (3.22)$$

By combining Eq. (3.19)~Eq. (3.22), the relationship between the time rate of change of B for the system and the control volume can be expressed as

联合式(3.19)~式(3.22)，基于系统和控制体的 B 的时间变化率之间的关系可以表示为

$$\frac{DB_{sys}}{Dt} = \frac{\partial B_{cv}}{\partial t} + \dot{B}_{out} - \dot{B}_{in} \quad (3.23)$$

This is a version of the Reynolds transport theorem. It can be seen that, since the inflow rate and the outflow rate of B for the control volume not always be the same, the time rate of change of B for the system [the left-hand side of Eq. (3.23)] is not necessarily the same as the rate of change of B within the control volume [the first term on the right-hand side of Eq. (3.23)].

这就是雷诺输运定理。由式可见，由于流入和流出控制体的广延参数 B 的速率不一定相同，系统的广延参数的时间变化率[式(3.23)左边部分]和控制体内的广延参数的时间变化率[式(3.23)右边第一项]不一定相同。

As is indicated in Fig. 3.15, in time δt the volume of fluid that passes across each area element is given by $\delta V = \delta l_n \delta A$, where $\delta l_n = \delta l \cos\theta$ is the height of the small volume element, and θ is the angle between the velocity vector and the unit vector in the direction of outer normal (**n**). Thus, since $\delta l = v\delta t$, the amount of B passes across the area element δA in the time interval δt is given by

如图 3.15 所示，δt 时间内流过每一个微元面的流体体积为 $\delta V = \delta l_n \delta A$，式中 $\delta l_n = \delta l \cos\theta$ 表示微元体的高度，θ 为速度矢量和外法线方向单位矢量 **n** 之间的夹角。由于 $\delta l = v\delta t$，δt 时间间隔内通过微元面 δA 的广延参数 B 的量为

$$\delta B = b\rho \delta V = b\rho(v\cos\theta \delta t)\delta A$$

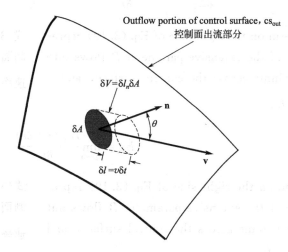

Fig. 3.15 Outflow across the control surface
图 3.15 通过控制面的出流

The rate of B that flows out of the control volume across the small area element δA denoted $\delta \dot{B}_{out}$ is

通过微元面 δA 流出控制体的广延参数 B 的速率 $\delta \dot{B}_{out}$ 为

$$\delta \dot{B}_{out} = \lim_{\delta t \to 0} \frac{\rho b \delta V}{\delta t} = \lim_{\delta t \to 0} \frac{(\rho b v \cos\theta \delta t)\delta A}{\delta t} = \rho b v \cos\theta \delta A$$

By integrating over the entire outflow portion of the control surface, it can be obtained that

对整个出流控制面进行积分，可得

$$\dot{B}_{out} = \int_{cs_{out}} d\dot{B}_{out} = \int_{cs_{out}} \rho b v \cos\theta dA$$

The quantity $v\cos\theta$ is the component of the velocity normal to the area element δA. According to the definition of the dot product, it can be written as $v\cos\theta = \mathbf{v} \cdot \mathbf{n}$. Hence, an alternate form of the outflow rate is

$v\cos\theta$ 为速度在微元面 δA 法线方向上的分量，根据点积的定义，可以表示为 $v\cos\theta = \mathbf{v} \cdot \mathbf{n}$。因此，$\dot{B}_{out}$ 的另一种形式为

$$\dot{B}_{out} = \int_{cs_{out}} \rho b \mathbf{v} \cdot \mathbf{n} dA \qquad (3.24)$$

In a similar method, by considering the inflow portion of the control surface cs_{in} as shown in Fig. 3.16, it can be obtained that the inflow rate of B into the control volume is

利用同样的方法，考虑如图 3.16 所示的入流面 cs_{in}，可得流入控制体的广延属性 B 的速率为

$$\dot{B}_{in} = -\int_{cs_{in}} \rho b v \cos\theta dA = -\int_{cs_{in}} \rho b \mathbf{v} \cdot \mathbf{n} dA \qquad (3.25)$$

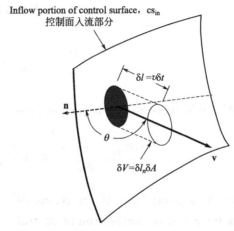

Fig. 3.16 Inflow across the control surface

图 3.16 通过控制面的入流

The unit normal vector to the control surface, $\mathbf{n}$, points out from the control volume. Thus, as shown in Fig. 3.17, in the outflow regions ($-90° < \theta < 90°$), the value of $\cos\theta$ is positive, and $\mathbf{v} \cdot \mathbf{n} > 0$. In the inflow regions ($90° < \theta < 270°$), the value of $\cos\theta$ is negative, and $\mathbf{v} \cdot \mathbf{n} < 0$. Over the remainder of the control surface, there is no inflow or outflow, and $\mathbf{v} \cdot \mathbf{n} = v\cos\theta = 0$. In such regions, either $v = 0$ (the fluid "sticks" to the surface) or $\cos\theta = 0$ (the fluid flows along the surface without crossing it). Therefore, the net flux of parameter B acrosses the entire control surface is

单位法向矢量 $\mathbf{n}$ 指向控制体外。因此，如图 3.17 所示，在出流区域（$-90° < \theta < 90°$），$\cos\theta$ 为正，$\mathbf{v} \cdot \mathbf{n} > 0$；在入流区域（$90° < \theta < 270°$），$\cos\theta$ 为负，$\mathbf{v} \cdot \mathbf{n} < 0$；控制面的其他区域没有入流或出流，$\mathbf{v} \cdot \mathbf{n} = v\cos\theta = 0$，在此区域，$v = 0$（流体黏附在表面上）或 $\cos\theta = 0$（流体流过表面而不穿透）。因此，通过整个控制面的广延参数 B 的净流率为

$$\dot{B}_{out} - \dot{B}_{in} = \int_{cs_{out}} \rho b \mathbf{v} \cdot \mathbf{n} dA - \left(-\int_{cs_{in}} \rho b \mathbf{v} \cdot \mathbf{n} dA\right) = \int_{cs} \rho b \mathbf{v} \cdot \mathbf{n} dA \qquad (3.26)$$

where the integration is over the entire control surface.

式（3.26）对整个控制面进行积分。

By combining Eq. (3.23) and Eq. (3.26), it can be obtained that

由式（3.23）和式（3.26）得

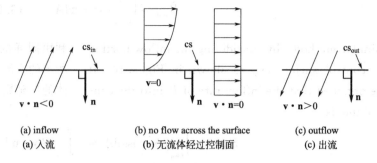

Fig. 3.17 Velocity configurations on the control surface
图 3.17 控制面上的速度分布

$$\frac{\mathrm{D}B_{\mathrm{sys}}}{\mathrm{D}t} = \frac{\partial B_{\mathrm{cv}}}{\partial t} + \int_{\mathrm{cs}} \rho b \mathbf{v} \cdot \mathbf{n} \mathrm{d}A$$

Since $B_{\mathrm{cv}} = \int_{\mathrm{cv}} \rho b \mathrm{d}V$, this can be written in another form

因为 $B_{\mathrm{cv}} = \int_{\mathrm{cv}} \rho b \mathrm{d}V$，上式以可写为

$$\frac{\mathrm{D}B_{\mathrm{sys}}}{\mathrm{D}t} = \frac{\partial}{\partial t}\int_{\mathrm{cv}} \rho b \mathrm{d}V + \int_{\mathrm{cs}} \rho b \mathbf{v} \cdot \mathbf{n} \mathrm{d}A \qquad (3.27)$$

Equation (3.27) is the general form of the Reynolds transport theorem for a fixed, nondeforming control volume. It is widely used in fluid mechanics and other areas as well.

式(3.27)为固定不变形控制体的雷诺输运定理的一般形式，它广泛应用于流体力学及其他领域。

3.5.2 Relationship Between Reynolds transport theorem and Material Derivative

3.5.2 雷诺输运定理与物质导数的关系

The material derivative introduced in section 3.2.1 holds the form of $\mathrm{D}(\ )/\mathrm{D}t = \partial(\ )/\partial t + \mathbf{v} \cdot \nabla(\ )$, its significance lies in that it provides the time rate of change of a fluid property (temperature, velocity, etc.) associated with a particular fluid particle as it flows. The value of that parameter for that particle may change due to unsteady effects [the $\partial(\ )/\partial t$ term] or effects associated with the particle's motion [the $\mathbf{v} \cdot \nabla(\ )$ term].

3.2.1 节中介绍的物质导数的形式为 $\mathrm{D}(\ )/\mathrm{D}t = \partial(\ )/\partial t + \mathbf{v} \cdot \nabla(\ )$，其意义在于给出了一个与特定流体质点相关的流体属性（温度、速度等）的时间变化率。其值可能因非定常作用 $[\partial(\ )/\partial t]$ 或质点运动的作用 $[\mathbf{v} \cdot \nabla(\ )]$ 而发生变化。

The Reynolds transport theorem [Eq. (3.27)] holds the same type of physical interpretation as the material derivative. The term involving the time derivative of the control volume integral represents unsteady effects,

雷诺输运定理 [式(3.27)] 和物质导数具有相同的物理意义。对控制体进行积分的时间导数项表示非定常作用，也就是说控制体内

that is to say values of the parameter within the control volume may change with time. The term involving the control surface integral represents the convective effects, which is associated with the flow of the system across the fixed control surface. The sum of these two terms gives the rate of change of the parameter B for the system. This corresponds to the interpretation of the material derivative.

Therefore, both the material derivative and the Reynolds transport theorem equations provide methods to transfer from the Lagrangian viewpoint to the Eulerian viewpoint. The material derivative [Eq. (3.5)] is essentially the infinitesimal (or derivative) equivalent of the finite size (or integral) Reynolds transport theorem [Eq. (3.27)].

Exercises

3.1 The velocity field of a flow is given by $\mathbf{v}=(y+2)\mathbf{i}+(3x-8)\mathbf{j}+4z\mathbf{k}$, where x, y and z are in meter. Determine the fluid speed at the origin ($x=y=z=0$) and the y axis ($x=z=0$).

3.2 The velocity field of a flow is given by $\mathbf{v}=4y/(x^2+y^2)^{1/2}\mathbf{i}-4x/(x^2+y^2)^{1/2}\mathbf{j}$, where x and y are in meter. Determine the fluid speed at the x axis and y axis. Determine the angles between the x axis and the velocity vector at points $(x, y)=(1, 0)$, $(1, 1)$ and $(0, 1)$, respectively.

3.3 The velocity field of a flow is given by $\mathbf{v}=-v_0 y/(x^2+y^2)^{1/2}\mathbf{i}+v_0 x/(x^2+y^2)^{1/2}\mathbf{j}$, where v_0 is a constant. Determine the equation of the streamlines and discuss the various characteristics of this flow.

3.4 A three-dimensional velocity field is given by $\mathbf{v}=x^2\mathbf{i}+2xy\mathbf{j}+y^2\mathbf{k}$. Determine the acceleration of this flow field.

3.5 The velocity of air in the diverging pipe shown in Fig. 3.18 is given by $v_1 = 8t$ m/s and $v_2 = 4t$ m/s, where t is in seconds. Determine (a) the local acceleration at points ① and ②; (b) the average convective acceleration between these two points.

3.5 如图 3.18 所示的扩散管内的速度为 $v_1 = 8t$ m/s, $v_2 = 4t$ m/s, 其中 t 的单位为秒。试确定:(a) 点①和②位置处的局部加速度;(b) 两点间的平均对流加速度。

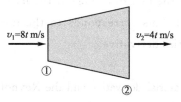

Fig. 3.18 Figure for the exercise 3.5
图 3.18 习题 3.5 图

3.6 Water flows through a constant diameter pipe with a uniform velocity given by $\mathbf{v} = (4/t + 2)\mathbf{j}$, where t is in seconds. Determine the acceleration at time $t = 1$, 3 and 5 s.

3.6 水以均匀的速度 $\mathbf{v} = (4/t + 2)\mathbf{j}$ 流过一个等径管,其中 t 的单位为秒。试确定 $t = 1$、3 和 5s 时刻的加速度。

3.7 A fluid flows along the x axis with a velocity given by $\mathbf{v} = (x/t)\mathbf{i}$, where x is in meter and t in seconds. (a) Plot the scattergram of speed for $0 \leqslant x \leqslant 8$m and $t = 4$s. (b) Plot the scattergram of speed for $x = 6$m and $1 \mathrm{s} \leqslant t \leqslant 3 \mathrm{s}$. (c) Determine the local and convective acceleration. (d) Show the locations that their acceleration of any fluid particle in the flow is zero. (e) Explain how the velocity of a particle in this unsteady flow remains constant throughout its motion.

3.7 流体以 $\mathbf{v} = (x/t)\mathbf{i}$ 的速度沿 x 轴流动,其中 x 的单位为米,t 的单位为秒。(a) 试画出 $0 \leqslant x \leqslant 8$m 范围内 $t = 4$s 时的速度分布图。(b) 试画出 $x = 6$m 位置处,$1 \mathrm{s} \leqslant t \leqslant 3 \mathrm{s}$ 范围内的速度分布图。(c) 试确定局部和对流加速度。(d) 标出流体质点加速度为零的位置。(e) 试解释在这种非定常流动中,质点运动的速度如何保持不变。

3.8 A fluid flows through a nozzle and the velocity increases from v_1 to v_2 in a linear fashion as shown in Fig. 3.19. That is, $v = ax + b$, where a and b are constants. If the flow is steady with $v_1 = 5$m/s at point ① ($x_1 = 0$) and $v_2 = 15$m/s at point ② ($x_2 = 1$m), determine the local acceleration, the convective acceleration, and the acceleration of the fluid at points ① and ②.

3.8 流体流经一个喷嘴,如图 3.19 所示,速度从 v_1 线性增加到 v_2,即 $v = ax + b$,其中 a 和 b 为常数。如果流动稳定,在点① ($x_1 = 0$) 处,速度 $v_1 = 5$m/s,在点② ($x_2 = 1$m) 处,速度 $v_2 = 15$m/s。试确定点①和②处的局部加速度、对流加速度以及流体的加速度。

3.9 If the flow in problem 3.8 is not steady, at the time when $v_1 = 5$m/s and $v_2 = 15$m/s, it is known that $\partial v_1 / \partial t = 20 \mathrm{m/s}^2$ and $\partial v_2 / \partial t = 40 \mathrm{m/s}^2$. Determine

3.9 如果 3.8 题中的流动是非定常的,某一时刻速度 $v_1 = 5$m/s, $v_2 = 15$m/s, 有 $\partial v_1 / \partial t = 20 \mathrm{m/s}^2$、

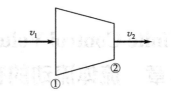

Fig. 3.19 Figure for the exercise 3.8
图 3.19 习题 3.8 图

the local acceleration, the convective acceleration, and the acceleration of the fluid at points ① and ②.

$\partial v_2/\partial t = 40 \text{m/s}^2$,试确定点①和②处的局部加速度、对流加速度以及流体的加速度。

3.10 The temperature of the exhaust in an exhaust pipe can be expressed as $T = T_0(1 + ae^{-bx})(1 + c\cos\omega t)$, where $T_0 = 120°C$, $a = 2$, $b = 0.02\text{m}^{-1}$, $c = 0.03$ and $\omega = 110\text{rad/s}$. If the exhaust speed is 3m/s, determine the time rate of change of temperature of the fluid particle at $x = 3\text{m}$ when $t = 0$.

3.10 排出管中废气的温度可表示为 $T = T_0(1+ae^{-bx})(1+c\cos\omega t)$,其中 $T_0 = 120°C$,$a = 2$,$b = 0.02\text{m}^{-1}$,$c = 0.03$,$\omega = 110\text{rad/s}$,如果排出速度为 3m/s,试确定当 $t = 0$ 时,$x = 3\text{m}$ 处流体质点的温度的时间变化率。

Chapter 4　Finite Control Volume Analysis of Fluid Flow
第4章　流体流动的有限控制体分析

Many fluid flow problems in engineering are analyzed based on a particular finite region (a control volume). In fluid mechanics, the control volume or Eulerian view is generally less complicated and more convenient to use than the system or Lagrangian view. In this chapter the finite control volume analysis of fluid flow will be introduced. This method is based on some fundamental principles of physics such as law of conservation of mass, Newton's second law of motion, and the first law of thermodynamics.

工程中的许多流体流动问题基于特定的有限区域（控制体）进行分析。在流体力学中，控制体法或欧拉法比系统法或拉格朗日法简单，更便于使用。本章将介绍流体流动的有限控制体分析，该方法基于物理学中的一些基本原理，如：质量守恒定律、牛顿第二运动定律和热力学第一定律。

4.1　The Continuity Equation

4.1　连续性方程

4.1.1　Derivation of the Continuity Equation

4.1.1　连续性方程的推导

In fluid mechanics, a system is defined as a collection of unchanging materials, for which the conservation of mass principle can be simply stated as

在流体力学中，系统定义为确定不变的物质集合，针对系统的质量守恒原理可简单表述为

$$\boxed{\text{time rate of change of the system mass}} = 0$$

$$\boxed{\text{系统质量的时间变化率}} = 0$$

or 或

$$\frac{\mathrm{D}M_{\text{sys}}}{\mathrm{D}t} = 0 \quad (4.1)$$

where the system mass, M_{sys}, can be expressed as

式中，系统的质量 M_{sys} 可以表达为

$$M_{\text{sys}} = \int_{\text{sys}} \rho \mathrm{d}V \quad (4.2)$$

As illustrated in Fig. 4.1, a system and a control volume are coincident at an instant of time. If $B=$ mass

如图4.1所示，系统和控制体瞬间重合，若 B 为质量，b 为1，则雷

and $b = 1$, the Reynolds transport theorem [Eq. (3.27)] can be expressed as

诺输运定理[式(3.27)]可表达为

$$\frac{D}{Dt}\int_{sys}\rho dV = \frac{\partial}{\partial t}\int_{cv}\rho dV + \int_{cs}\rho \mathbf{v} \cdot \mathbf{n} dA \quad (4.3)$$

or 或

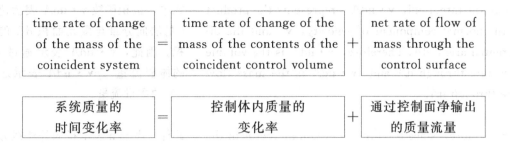

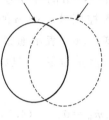 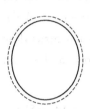 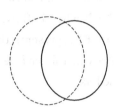

(a) System and control volume at time $t-\delta t$.
(a) $t-\delta t$ 时刻的系统和控制体

(b) System and control volume at time t, coincident condition.
(b) t 时刻的系统和控制体，两者重合

(c) System and control volume at time $t+\delta t$.
(c) $t+\delta t$ 时刻的系统和控制体

Fig. 4.1 System and control volume at three different instances of time
图 4.1 三个不同时刻的系统和控制体

The control volume expression for conservation of mass is commonly called the continuity equation. For a control volume, the combination of Eq. (4.1), Eq. (4.2) and Eq. (4.3) gives

质量守恒的控制体表达式通常称为连续性方程，对于控制体，联合式(4.1)、式(4.2)和式(4.3)可得

$$\frac{\partial}{\partial t}\int_{cv}\rho dV + \int_{cs}\rho \mathbf{v} \cdot \mathbf{n} dA = 0 \quad (4.4)$$

Equation (4.4) indicates that the summation of the time rate of change of the mass in control volume and the net mass flow rate through the control surface is equal to zero.

式(4.4)表示控制体内的质量变化率加上通过控制面的净质量流量等于零。

For steady flow, the time rate of change of the mass in control volume is zero. That is

对于定常流动，控制体内质量的时间变化率为零，即

$$\frac{\partial}{\partial t}\int_{cv}\rho \mathrm{d}V = 0 \quad (4.5)$$

Therefore, Eq. (4.4) can be written as

$$\int_{cs}\rho \mathbf{v}\cdot\mathbf{n}\mathrm{d}A = 0 \quad (4.6)$$

where the integrand, $\mathbf{v}\cdot\mathbf{n}\mathrm{d}A$, represents the product of the normal component of velocity, $\mathbf{v}$, and the differential area, $\mathrm{d}A$. Therefore, $\mathbf{v}\cdot\mathbf{n}\mathrm{d}A$ is the volume flow rate through $\mathrm{d}A$ and $\rho\mathbf{v}\cdot\mathbf{n}\mathrm{d}A$ is the mass flow rate through $\mathrm{d}A$.

Furthermore, since $\mathbf{n}$ represents the unit vector in the exterior normal direction, the sign of the dot product $\mathbf{v}\cdot\mathbf{n}$ is "$+$" when fluid flows out of the control volume and is "$-$" when it flows into the control volume. Equation(4.6) represents that the net mass flow rate through the control surface is zero, and it can also be expressed as

$$\int_{cs}\rho\mathbf{v}\cdot\mathbf{n}\mathrm{d}A = \sum\dot{m}_{\text{out}} - \sum\dot{m}_{\text{in}} = 0 \quad (4.7)$$

where $\dot{m}$ is the mass flow rate.

An frequently-used expression for mass flow rate through a section of control surface with area A is

$$\dot{m} = \rho Q = \rho A \bar{v} \quad (4.8)$$

where ρ is the average density of fluid, $\bar{v}$ is the average normal velocity with respect to the area A, and Q is the volume flow rate. The fluid velocity used in Eq. (4.8) is defined as

$$\bar{v} = \frac{\int_A \rho \mathbf{v}\cdot\mathbf{n}\mathrm{d}A}{\rho A} \quad (4.9)$$

If the velocity is uniformly distributed over the section area, A, then

$$\bar{v} = \frac{\int_A \rho \mathbf{v}\cdot\mathbf{n}\mathrm{d}A}{\rho A} = v \quad (4.10)$$

and the notation "—" over the v can be omitted. When the flow is not uniformly distributed over the flow cross-sectional area, the notation "—" reminds us that an average velocity is used.

If the steady flow is also incompressible, the net volume flow rate through the control surface is also zero.

$$\sum Q_{out} - \sum Q_{in} = 0 \qquad (4.11)$$

For a steady flow flowing through the control volume at section 1 and section 2

$$\dot{m} = \rho_1 A_1 \bar{v}_1 = \rho_2 A_2 \bar{v}_2 \qquad (4.12)$$

and if it is incompressible

$$Q = A_1 \bar{v}_1 = A_2 \bar{v}_2 \qquad (4.13)$$

4.1.2 Application of the Continuity Equation

Two application examples of the continuity equation are as follows.

【EXAMPLE 4.1】 As shown in Fig. 4.2, the water flows steadily through a nozzle at the end of a fire hose connected with a pump. If the nozzle exit velocity is 25m/s, determine the flow rate of the pump.

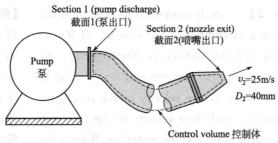

Fig. 4.2 Figure for the example 4.1

SOLUTION The pump output is the volume flow rate discharged from the pump to the nozzle. The control volume is designated with the dashed line in Fig. 4.2. By applying Eq. (4.4) to the control volume, we have

$$\frac{\partial}{\partial t}\int_{cv}\rho dV + \int_{cs}\rho \mathbf{v}\cdot\mathbf{n}dA = 0 \quad (4.14)$$

Since it is a steady flow, the time rate of change of the mass in this control volume is zero. According to Eq. (4.7), the total mass flow rate at section 1 and section 2 can be expressed as

$$\int_{cs}\rho\mathbf{v}\cdot\mathbf{n}dA = \dot{m}_2 - \dot{m}_1 = 0$$

Therefore

$$\dot{m}_2 = \dot{m}_1 \quad (4.15)$$

Since the mass flow rate is equal to the product of fluid density, ρ, and volume flow rate, Q, it can be obtained from Eq. (4.15) that

$$\rho_2 Q_2 = \rho_1 Q_1 \quad (4.16)$$

In this example, water can be considered incompressible. Therefore $\rho_2 = \rho_1$ and from Eq. (4.16)

$$Q_2 = Q_1 \quad (4.17)$$

The pump output is equal to the volume flow rate at the nozzle exit. For simplicity, the velocity distribution at the nozzle exit, section 2, is considered uniform, then from Eq. (4.17), Eq. (4.8) and Eq. (4.10)

$$Q_1 = Q_2 = v_2 A_2 = v_2 \frac{\pi}{4}D_2^2 = 25\times\frac{\pi}{4}\times 0.04^2 = 0.0314 \text{ (m}^3\text{/s)}$$

【EXAMPLE 4.2】 An incompressible and laminar flow develops in a straight pipe with radius of R, as indicated in Fig. 4.3. At section 1, the velocity, v, distribution is assumed to be uniform and it is parallel to the pipe axis. At section 2, the velocity profile is axisymmetric and parabolic, with zero velocity at the pipe wall and a maximum value of v_{max} at the centerline. What is the relationship between v and v_{max}? How are the average velocity at section 2, $\bar{v}_2$ and v_{max}, related?

SOLUTION A control volume sketched by dashed lines in Fig. 4.3 is selected. The application of Eq. (4.4) to the control volume yields

由于流入和流出控制体的流动是定常流动,控制体内的质量变化率为零。根据式(4.7),流经截面1和2的质量流量可表达为

因此

由于质量流量等于密度ρ与体积流量Q的乘积,由式(4.15)可得

本例中,水可当作不可压缩流体,因此$\rho_2 = \rho_1$,由式(4.16)得

泵的出流量等于喷嘴出口的体积流量。为简便起见,认为喷嘴出口(截面2)上流速均匀分布,由式(4.17)、式(4.8)和式(4.10)得

【例题4.2】 半径为R的圆直管中的不可压缩层流流动,如图4.3所示。截面1处,速度v均匀分布,且平行于管轴线。截面2处,速度呈轴对称抛物线型分布,管壁处速度为零,最大速度v_{max}出现在管中心处。请回答:v和v_{max}之间的关系是什么?截面2处的平均速度$\bar{v}_2$和v_{max}的关系是什么?

解 选取图4.3中虚线所示的控制体,将式(4.4)应用于此控制体得

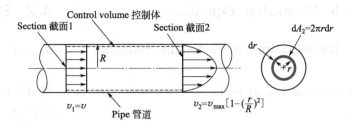

Fig. 4.3 Figure for the example 4.2
图 4.3 例题 4.2 图

$$\int_{cs} \rho \mathbf{v} \cdot \mathbf{n} dA = 0$$

The surface integral is evaluated at section 1 and section 2 to give

在截面 1 和截面 2 处面积分得

$$-\rho_1 A_1 v + \int_{A_2} \rho_2 \mathbf{v} \cdot \mathbf{n} dA_2 = 0 \quad (4.18)$$

At section 2, $\mathbf{v} \cdot \mathbf{n} = v_2$ and $dA_2 = 2\pi r dr$, so Eq. (4.18) becomes

在截面 2 上，$\mathbf{v} \cdot \mathbf{n} = v_2$，$dA_2 = 2\pi r dr$，式(4.18) 变为

$$-\rho_1 A_1 v + \rho_2 \int_0^R v_2 2\pi r dr = 0 \quad (4.19)$$

Considering $\rho_1 = \rho_2$ and using the parabolic velocity relationship at section 2, Eq. (4.19) is changed into the form

考虑到 $\rho_1 = \rho_2$，并将截面 2 上的抛物线速度分布代入式(4.19) 得

$$-A_1 v + 2\pi v_{max} \int_0^R \left[1 - \left(\frac{r}{R}\right)^2\right] r dr = 0 \quad (4.20)$$

Integrating, it can be obtained from Eq. (4.20) that

对式(4.20) 进行积分得

$$-\pi R^2 v + 2\pi v_{max} \left(\frac{r^2}{2} - \frac{r^4}{4R^2}\right)\bigg|_0^R = 0$$

$$v_{max} = 2v$$

Since this flow is incompressible, it can be concluded from Eq. (4.10) that v is the average velocity at all sections of the control volume. Thus, the average velocity at section 2, $\bar{v}_2$, is equal to v. Therefore

由于内流动是不可压缩流动，由式(4.10) 得，v 即为控制体各截面上的平均速度。因此，截面 2 上的平均速度 $\bar{v}_2$ 等于 v。故

$$\bar{v}_2 = \frac{v_{max}}{2}$$

4.2 The Momentum Equation

4.2.1 Derivation of the Momentum Equation

Newton's second law of motion for a system can be stated as

| time rate of change of the momentum of the system | = | sum of external forces acting on the system |

Since momentum is the product of mass and velocity, the momentum of a small particle with mass ρdV is $\mathbf{v}\rho dV$. Thus, the momentum of the entire system is $\int_{sys} \mathbf{v}\rho dV$ and the Newton's second law can be expressed as

$$\frac{D}{Dt}\int_{sys} \mathbf{v}\rho dV = \sum \mathbf{F}_{sys} \qquad (4.21)$$

When a control volume is coincident with a system at an instant of time (see Fig. 4.4), the forces acting on the system and that acting on the contents of the coincident control volume are instantaneously identical, that is

$$\sum \mathbf{F}_{sys} = \sum \mathbf{F}_{cv} \qquad (4.22)$$

Furthermore, for a system and a coincident control volume, if b is equal to the velocity and B_{sys} is the system momentum, the Reynolds transport theorem [Eq. (3.27)] can be expressed as

$$\frac{D}{Dt}\int_{sys} \mathbf{v}\rho dV = \frac{\partial}{\partial t}\int_{cv} \mathbf{v}\rho dV + \int_{cs} \mathbf{v}\rho \mathbf{v} \cdot \mathbf{n} dA \qquad (4.23)$$

or

| time rate of change of the momentum of the system | = | time rate of change of the momentum of the contents of the control volume | + | net rate of flow of the momentum through the control surface |

4.2 动量方程

4.2.1 动量方程的推导

针对系统的牛顿第二运动定律可表述为

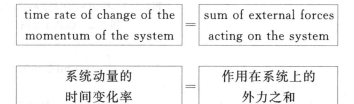

因为动量是质量与速度的乘积,所以质量为 ρdV 的微小颗粒的动量为 $\mathbf{v}\rho dV$,整个系统的动量为 $\int_{sys} \mathbf{v}\rho dV$,牛顿第二运动定律可表达为

当控制体和系统瞬间重合时(参见图 4.4),作用在系统上的力和作用在控制体上的力瞬间一致,即

此外,对于系统和与之重合的控制体,若 b 等于速度,B_{sys} 为系统的动量,则雷诺输运定理[式(3.27)]可表达为

或

| 系统动量的时间变化率 | = | 控制体内动量的时间变化率 | + | 通过控制面净输出的动量流量 |

Combination of Eq. (4.21)~Eq. (4.23) gives 联合式(4.21)~式(4.23)，得

$$\frac{\partial}{\partial t}\int_{cv}\mathbf{v}\rho dV + \int_{cs}\mathbf{v}\rho\mathbf{v}\cdot\mathbf{n}dA = \sum\mathbf{F}_{cv} \quad (4.24)$$

Equation (4.24) is called the momentum equation. 式(4.24)称作动量方程。

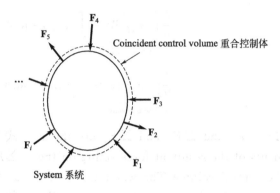

Fig. 4.4　External forces acting on system and coincident control volume
图 4.4　作用在系统和重合控制体上的外力

4.2.2　Application of the Momentum Equation

4.2.2　动量方程的应用

The momentum equation is a vector equation. In applications, it will be more convenient to use any component of this vector equation resolved along orthogonal coordinates, for example, x, y and z in rectangular coordinate system.

动量方程为矢量方程，应用过程中，使用沿正交坐标轴分解的分量方程更为方便，如：直角坐标系中沿 x、y 和 z 分解。

【EXAMPLE 4.3】　As shown in Fig. 4.5(a), a horizontal stream of water with a uniform speed of $v_1 = 3\text{m/s}$ impacts a slope, and is turned through an angle $\theta = 45°$. Gravity and viscous effects can be ignored. Determine the fixing force exerting on the slope.

【例题 4.3】　如图 4.5(a) 所示，水平水流以 3m/s 的速度冲击斜坡，然后 45°转弯。忽略重力和黏性力，试确定作用在斜坡上的固定力。

SOLUTION　The control volume in this problem is selected as shown in Fig. 4.5 (b). The x and z components of the momentum equation Eq. (4.24) can be expressed as

解　选择的控制体如图 4.5(b) 所示。则式(4.24) 的 x 和 z 方向的分量可表达为

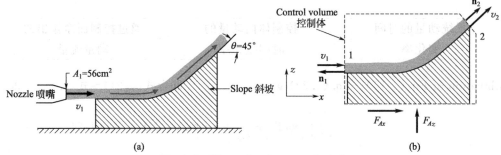

Fig. 4.5 Figure for the example 4.3

图 4.5 例题 4.3 图

$$\underbrace{\frac{\partial}{\partial t}\int_{cv} v_x \rho \mathrm{d}V}_{=0(\text{flow is steady})} + \int_{cs} v_x \rho \mathbf{v} \cdot \mathbf{n} \mathrm{d}A = \sum F_x \quad (4.25\mathrm{a})$$

$$\underbrace{\frac{\partial}{\partial t}\int_{cv} v_z \rho \mathrm{d}V}_{=0(\text{flow is steady})} + \int_{cs} v_z \rho \mathbf{v} \cdot \mathbf{n} \mathrm{d}A = \sum F_z \quad (4.25\mathrm{b})$$

where $\mathbf{v}=v_x\mathbf{i}+v_z\mathbf{k}$, and $\sum F_x$ and $\sum F_z$ are the x and z components of the resultant force acting on the contents of the control volume. The water flows into and out of the control volume as a free stream. The control volume is surrounded by the atmospheric pressure, so the net pressure force on the control volume surface is zero. Since the gravity is neglected, the only forces applied to the control volume contents are the horizontal and vertical components of the fixing force, F_{Ax} and F_{Az}, respectively.

At section 1, $\mathbf{v}\cdot\mathbf{n}=-v_1$, and at section 2, $\mathbf{v}\cdot\mathbf{n}=+v_2$. Also, with negligible gravity and viscous effects, and since $p_1=p_2$, the speed of the fluid remains constant, so that $v_1=v_2=3\mathrm{m/s}$. Hence, at section 1, $v_x=v_1$, $v_z=0$, and at section 2, $v_x=v_1\cos\theta$, $v_z=v_1\sin\theta$. According to the above analysis, Eq. (4.25) can be written as

式中，$\mathbf{v}=v_x\mathbf{i}+v_z\mathbf{k}$，$\sum F_x$ 和 $\sum F_z$ 为作用在控制体上的合力在 x 和 z 方向的分量。水以自由射流的形式进入和离开控制体，控制体周围的压力为大气压，因此，作用在控制体上的净压力为零。由于忽略重力，因此，作用在控制体上的力分别是水平方向的力 F_{Ax} 和竖直方向的力 F_{Az}。

在截面 1 处，$\mathbf{v}\cdot\mathbf{n}=-v_1$；在截面 2 处，$\mathbf{v}\cdot\mathbf{n}=+v_2$。由于忽略重力和黏性作用，并且 $p_1=p_2$，故流速保持不变，所以 $v_1=v_2=3\mathrm{m/s}$。因此，在截面 1 处，$v_x=v_1$，$v_z=0$；在截面 2 处，$v_x=v_1\cos\theta$，$v_z=v_1\sin\theta$。根据上述分析，式(4.25) 可写为

$$v_1\rho(-v_1)A_1+v_1\cos\theta\rho(v_1)A_2=F_{Ax} \quad (4.26\mathrm{a})$$

$$(0)\rho(-v_1)A_1+v_1\sin\theta\rho(v_1)A_2=F_{Az} \quad (4.26\mathrm{b})$$

For this incompressible flow, according to conservation of mass, $A_1v_1=A_2v_2$, or $A_1=A_2$ since $v_1=v_2$. Thus, Eq. (4.26) can be simplified as

对于不可压缩流动，根据质量守恒，$A_1v_1=A_2v_2$，由于 $v_1=v_2$，则 $A_1=A_2$。因此，式(4.26) 可简化为

$$F_{Ax} = -\rho A_1 v_1^2 + \rho A_1 v_1^2 \cos\theta = -\rho A_1 v_1^2 (1-\cos\theta)$$

$$F_{Az} = \rho A_1 v_1^2 \sin\theta$$

With the given data, it can be obtained that 代入数据，得

$$F_{Ax} = -1000 \times 0.0056 \times 3^2 \times (1-\cos 45°) = -14.76 \text{ (N)}$$

$$F_{Az} = 1000 \times 0.0056 \times 3^2 \times \sin 45° = 35.63 \text{ (N)}$$

4.3 Moment-of-Momentum Equation

4.3.1 Derivation of the Moment-of-Momentum Equation

The moment of a force with respect to an axis is called torque. When torques should be considered in a problem, the moment-of-momentum equation is often more convenient to use than the momentum equation.

Application of Newton's second law of motion to a particle of fluid yields

4.3 动量矩方程

4.3.1 动量矩方程的推导

力相对于某一轴的矩称作为扭矩。当研究的问题中需要考虑扭矩时，采用动量矩方程比动量方程更方便。

对流体微团应用牛顿第二运动定律，有

$$\frac{D}{Dt}(\mathbf{v}\rho\delta V) = \delta \mathbf{F}_p \qquad (4.27)$$

where $\mathbf{v}$ is the particle velocity, ρ is the particle density, δV is the particle volume, and $\delta \mathbf{F}_p$ is the resultant force acting on the particle. Multiplied by the vector, $\mathbf{r}$, in each side of Eq. (4.27), it can be obtained that

式中，$\mathbf{v}$ 是微团的速度；ρ 为微团的密度；δV 为微团的体积；$\delta \mathbf{F}_p$ 为作用在微团上的合力。式 (4.27) 的两边乘以矢径 $\mathbf{r}$，可得

$$\mathbf{r} \times \frac{D}{Dt}(\mathbf{v}\rho\delta V) = \mathbf{r} \times \delta \mathbf{F}_p \qquad (4.28)$$

where $\mathbf{r}$ is the position vector from the origin of the inertial coordinate system to the fluid particle (see Fig. 4.6). It can be noted that

式中，$\mathbf{r}$ 为惯性坐标系原点指向流体微团的位置矢径（参见图 4.6）。注意到

$$\frac{D}{Dt}[(\mathbf{r}\times\mathbf{v})\rho\delta V] = \frac{D\mathbf{r}}{Dt}\times\mathbf{v}\rho\delta V + \mathbf{r}\times\frac{D(\mathbf{v}\delta V)}{Dt} \qquad (4.29)$$

and 且

$$\frac{D\mathbf{r}}{Dt} = \mathbf{v} \qquad (4.30)$$

Thus, since 因此，有

$$\mathbf{v} \times \mathbf{v} = 0 \quad (4.31)$$

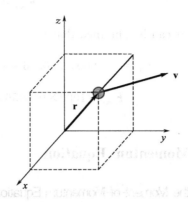

Fig. 4.6 Inertial coordinate system
图 4.6 惯性坐标系

by combining Eq. (4.28)~Eq. (4.31), it can be obtained that 由式(4.28)~式(4.31)，可得

$$\frac{D}{Dt}[(\mathbf{r} \times \mathbf{v})\rho \delta V] = \mathbf{r} \times \delta \mathbf{F}_p \quad (4.32)$$

Equation (4.32) is valid for every particle in a system. For a system (collection of fluid particles), by using the sum of both sides of Eq. (4.32), it can be obtained that

式(4.32)对系统中的所有微团均成立。对于系统（即流体微团的集合），式(4.32)两边求和可得

$$\int_{sys} \frac{D}{Dt}[(\mathbf{r} \times \mathbf{v})\rho dV] = \sum (\mathbf{r} \times \mathbf{F})_{sys} \quad (4.33)$$

where 其中

$$\sum (\mathbf{r} \times \mathbf{F})_{sys} = \sum \mathbf{r} \times \delta \mathbf{F}_p \quad (4.34)$$

Since the sequential order of differential and integral has no effect on the result, so

由于微分和积分的顺序对换对结果没有影响，故

$$\frac{D}{Dt}\int_{sys} (\mathbf{r} \times \mathbf{v})\rho dV = \int_{sys} \frac{D}{Dt}[(\mathbf{r} \times \mathbf{v})\rho dV] \quad (4.35)$$

Thus, Eq. (4.33) can be expressed as 因此，式(4.33)可写为

$$\frac{D}{Dt}\int_{sys} (\mathbf{r} \times \mathbf{v})\rho dV = \sum (\mathbf{r} \times \mathbf{F})_{sys} \quad (4.36)$$

or 或

$$\begin{bmatrix} \text{time rate of change of the} \\ \text{moment-of-momentum of the system} \end{bmatrix} = \begin{bmatrix} \text{sum of external torques} \\ \text{acting on the system} \end{bmatrix}$$

| 系统动量矩的
时间变化率 | = | 作用在系统上的
外力矩之和 |

When the control volume is instantaneously coincident with the system, the torques acting on the system will be identical to that on the control volume contents. That is

当控制体和系统瞬间重合时，作用在系统上的扭矩和作用在控制体上的扭矩一致，即

$$\sum(\mathbf{r}\times\mathbf{F})_{sys}=\sum(\mathbf{r}\times\mathbf{F})_{cv} \quad (4.37)$$

In this case, if b is the velocity moment and B_{sys} is the moment-of-momentum of the system, the Reynolds transport theorem [Eq. (3.27)] can be expressed as

这种情况下，若 b 为速度矩，B_{sys} 为系统的动量矩，则雷诺输运定理 [式(3.27)] 可表达为

$$\frac{D}{Dt}\int_{sys}(\mathbf{r}\times\mathbf{v})\rho dV=\frac{\partial}{\partial t}\int_{cv}(\mathbf{r}\times\mathbf{v})\rho dV+\int_{cs}(\mathbf{r}\times\mathbf{v})\rho\mathbf{v}\cdot\mathbf{n}dA \quad (4.38)$$

or

或

| time rate of change of the
moment-of-momentum
of the system | = | time rate of change of the moment-of-momentum of the contents of the coincident control volume | + | net rate of flow of the
moment-of-momentum
through the control surface |

| 系统动量矩的
时间变化率 | = | 控制体内动量矩的
时间变化率 | + | 通过控制面净输出的
动量矩流量 |

By combining Eq. (4.36)~Eq. (4.38), the moment-of-momentum equation can be expressed as

联合式(4.36)~式(4.38)，动量矩方程可表达为

$$\frac{\partial}{\partial t}\int_{cv}(\mathbf{r}\times\mathbf{v})\rho dV+\int_{cs}(\mathbf{r}\times\mathbf{v})\rho\mathbf{v}\cdot\mathbf{n}dA=\sum(\mathbf{r}\times\mathbf{F})_{cv} \quad (4.39)$$

The moment-of-momentum equation [Eq. (4.39)] is often used to solve the fluid mechanical problems in the rotating machines, such as rotary lawn sprinklers, ceiling fans, wind turbines and centrifugal pumps.

动量矩方程通常用于求解旋转机械中的流体力学问题，如：旋转式草坪喷洒器、吊扇、风力透平机以及离心泵。

4.3.2 Application of the Moment-of-Momentum Equation

4.3.2 动量矩方程的应用

As shown in Fig. 4.7, water flows through a sprinkler from the inlet (section 1) to the outlet (section 2) and during the course a torque is generated on the sprinkler head causing it to rotate.

如图 4.7 所示，水从喷洒器的进口（截面1）流到出口（截面2），在此过程中，在喷头上产生了一个使其旋转的扭矩。

The moment-of-momentum equation [Eq. (4.39)] can be applied on the control volume shown in Fig. 4.7. When the sprinkler is rotating, the flow field in the stationary control volume is cyclical and unsteady, but steady in the mean. The axial component of the moment-of-momentum equation [Eq. (4.39)] can be used to analyze this flow.

动量矩方程[式(4.39)]可应用于如图 4.7 所示的控制体。当喷洒器旋转时,控制体内的流场是周期、非定常的,但时间平均以后可视为定常流动。式(4.39)中的轴向分量方程可对此流动进行分析。

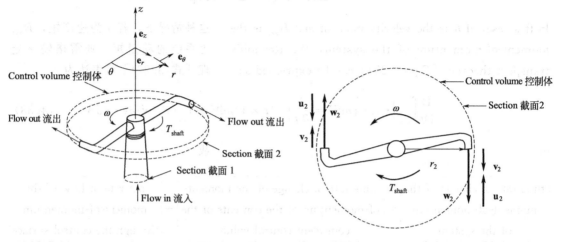

Fig. 4.7 Rotary water sprinkler

图 4.7 旋转喷洒器

At section 1, water flows into the control volume axially. At this position, the direction of $\mathbf{r} \times \mathbf{v}$ is perpendicular to the axis of rotation, so the component of $\mathbf{r} \times \mathbf{v}$ along the axis of rotation is zero. Thus, there is no axial moment-of-momentum flow in at section 1.

在截面 1 处水沿轴向进入控制体,此处,$\mathbf{r} \times \mathbf{v}$ 的方向垂直于旋转轴,所以 $\mathbf{r} \times \mathbf{v}$ 沿旋转轴的分量为零,因此,在截面 1 处没有轴向动量矩流量。

At section 2, water leaves the control volume. At this position, the magnitude of the axial component of $\mathbf{r} \times \mathbf{v}$ is $r_2 v_{\theta 2}$, where r_2 is the radius from the axis of rotation to the nozzle centerline and $v_{\theta 2}$ is the value of the tangential component of the absolute velocity of the exiting flow.

在截面 2 处水流出控制体,此处,$\mathbf{r} \times \mathbf{v}$ 的轴向分量的大小为 $r_2 v_{\theta 2}$,r_2 为旋转轴到喷嘴中心线的半径,$v_{\theta 2}$ 为绝对出流速度的切向分量。

The velocity of outflow relative to the fixed control surface is absolute velocity, $\mathbf{v}$, and the velocity of outflow relative to the nozzle is relative velocity, $\mathbf{w}$, then the relationship between $\mathbf{v}$ and $\mathbf{w}$ can be expressed as

相对于固定控制面的出流速度为绝对速度 $\mathbf{v}$。相对于喷嘴的出流速度为相对速度 $\mathbf{w}$。那么,$\mathbf{v}$ 和 $\mathbf{w}$ 之间的关系可表达为

$$\mathbf{v} = \mathbf{w} + \mathbf{u} \quad (4.40)$$

where **u** is the velocity of the moving nozzle relative to the fixed control surface.

式中，**u** 为相对于固定控制面的喷嘴运动速度。

The dot product $\mathbf{v} \cdot \mathbf{n}$ and cross product $\mathbf{r} \times \mathbf{v}$ involved in Eq. (4.39) can be positive or negative. For flow into the control volume, $\mathbf{v} \cdot \mathbf{n}$ is negative. For flow out, $\mathbf{v} \cdot \mathbf{n}$ is positive. The algebraic sign and direction of $\mathbf{r} \times \mathbf{v}$ can be ascertained by using the right-hand rule. Thus, the axial component of moment-of-momentum flow can be written as

式(4.39) 中的叉积和点积可以为正值或负值。流体流进控制体，$\mathbf{v} \cdot \mathbf{n}$ 为负值；流出控制体，$\mathbf{v} \cdot \mathbf{n}$ 为正值。$\mathbf{r} \times \mathbf{v}$ 的正负和方向可以通过右手法则确定。因此，动量矩流量的轴向分量可写为

$$\left[\int_{cs} (\mathbf{r} \times \mathbf{v}) \rho \mathbf{v} \cdot \mathbf{n} dA \right]_{axial} = (-r_2 v_{\theta 2})(+\dot{m}) \quad (4.41)$$

where $\dot{m}$ is the total mass flow rate through the two nozzles. The algebraic sign of the axial component of $\mathbf{r} \times \mathbf{v}$ can be easily judged according to the rule: if v_θ and **u** are in the same direction, use "+"; if v_θ and **u** are in opposite directions, use "−".

式中，$\dot{m}$ 为通过两个喷嘴的总质量流量。$\mathbf{r} \times \mathbf{v}$ 的正负可按下面方法简单判断：如果，v_θ 和 **u** 的方向相同，为正；如果 v_θ 和 **u** 的方向相反，为负。

For the sprinkler shown in Fig. 4.7, the torque term $[\Sigma(\mathbf{r} \times \mathbf{v})_{cv}]$ of the moment-of-momentum equation [Eq. (4.39)] can be written as

对于如图 4.7 所示的喷洒器，动量矩方程 [式(4.39)] 的扭矩项 $\Sigma(\mathbf{r} \times \mathbf{v})_{cv}$ 可写为

$$T_{shaft} = \Sigma [(\mathbf{r} \times \mathbf{F})_{cv}]_{axial} \quad (4.42)$$

where T_{shaft} is positive by assuming that T_{shaft} is in the same direction as rotation.

式中，假设 T_{shaft} 的方向与旋转方向相同，T_{shaft} 为正值。

From Eq. (4.41) and Eq. (4.42), the axial component of the moment-of-momentum equation [Eq. (4.39)] is

由式(4.41) 和式(4.42) 得动量矩方程 [式(4.39)] 的轴向分量为

$$-r_2 v_{\theta 2} \dot{m} = T_{shaft} \quad (4.43)$$

The shaft power, $\dot{W}_{shaft}$, is equal to the product of T_{shaft} and the rotational speed of the shaft, ω. Thus, from Eq. (4.43), it can be obtained that

轴功率 $\dot{W}_{shaft}$ 等于轴扭矩 T_{shaft} 与轴转速 ω 的乘积，因此，由式(4.43) 得

$$\dot{W}_{shaft} = T_{shaft} \omega = -r_2 v_{\theta 2} \dot{m} \omega \quad (4.44)$$

Since $r_2 \omega$ is the speed of each sprinkler nozzle, u_2, Eq. (4.44) can also be written as

由于 $r_2 \omega$ 等于喷洒器喷嘴的速度 u_2，式(4.44) 也可表达为

$$\dot{W}_{shaft} = -u_2 v_{\theta 2} \dot{m} \qquad (4.45)$$

Shaft work per unit mass, w_{shaft}, is equal to $\dot{W}_{shaft}/\dot{m}$. Dividing Eq. (4.45) by the mass flow rate, $\dot{m}$, it can be obtained that

$$w_{shaft} = -u_2 v_{\theta 2} \qquad (4.46)$$

The negative sign in Eq. (4.44)~Eq. (4.46) represents the work out of the control volume.

4.4 The Energy Equation

4.4.1 Derivation of the Energy Equation

The first law of thermodynamics for a system can be expressed as

time rate of increase of the total stored energy of the system	=	net time rate of energy addition by heat transfer into the system	+	net time rate of energy addition by work transfer into the system
系统总储存能量的增加率	=	以传热形式净输入系统的能量增加率	+	以做功形式净输入系统的能量增加率

In symbolic form

$$\frac{D}{Dt}\int_{sys} e\rho\, dV = \left(\sum \dot{Q}_{in} - \sum \dot{Q}_{out}\right)_{sys} + \left(\sum \dot{W}_{in} - \sum \dot{W}_{out}\right)_{sys}$$

or

$$\frac{D}{Dt}\int_{sys} e\rho\, dV = (\dot{Q}_{net\,in} + \dot{W}_{net\,in})_{sys} \qquad (4.47)$$

The total stored energy per unit mass, e, includes the internal energy per unit mass, u, the kinetic energy per unit mass, $v^2/2$, and the potential energy per unit mass, gz. It can be expressed as

$$e = u + \frac{v^2}{2} + gz \qquad (4.48)$$

The net rate of heat transfer into the system is labeled $\dot{Q}_{net\,in}$, and the net rate of work transfer into the system is labeled $\dot{W}_{\text{net in}}$. Heat and work transfer into the system are considered positive and transfer out are negative.

When the control volume is coincident with the system at an instant of time

$$(\dot{Q}_{\text{net in}} + \dot{W}_{\text{net in}})_{\text{sys}} = (\dot{Q}_{\text{net in}} + \dot{W}_{\text{net in}})_{\text{cv}} \quad (4.49)$$

Furthermore, if b set equal to e, the Reynolds transport theorem [Eq. (3.27)] can be expressed as

$$\frac{\text{D}}{\text{D}t}\int_{\text{sys}} e\rho \text{d}V = \frac{\partial}{\partial t}\int_{\text{cv}} e\rho \text{d}V + \int_{\text{cs}} e\rho \mathbf{v} \cdot \mathbf{n} \text{d}A \quad (4.50)$$

or in words

the time rate of increase of the total stored energy of the system	=	the time rate of increase of the toral stored energy of the contents of the control volume	+	the net rate of flow of the total stored energy out of the control volume through the control surface
系统总储存能量的增加率	=	控制体内总储存能量的增加率	+	通过控制面净输出控制体的总储存能量流量

Combining Eq. (4.47), Eq. (4.49) and Eq. (4.50), the first law of thermodynamics for the control volume can be written as

$$\frac{\partial}{\partial t}\int_{\text{cv}} e\rho \text{d}V + \int_{\text{cs}} e\rho \mathbf{v} \cdot \mathbf{n} \text{d}A = (\dot{Q}_{\text{net in}} + \dot{W}_{\text{net in}})_{\text{cv}} \quad (4.51)$$

The heat transfer rate, $\dot{Q}$, represents the energy exchange per unit time between the control volume contents and surroundings in the ways of radiation, conduction, or convection.

The work transfer rate, $\dot{W}$, also called power, is positive when work is done on the contents of the control volume by the surroundings. Otherwise, it is considered negative. Some important forms of power transfer include the net in shaft power $\dot{W}_{\text{shaft net in}}$, the normal stress force power $\dot{W}_{\text{ns}}$ and the shear stress force power, $\dot{W}_{\text{ts}}$. In general, shear stress force power is zero or negligibly small.

当控制体和系统瞬间一致时，有

此外，若 b 等于 e，则雷诺输运定理 [式(3.27)] 可表达为

用文字表达为

联立式(4.47)、式(4.49)和式(4.50)，基于控制体的热力学第一定律可表达为

传热速率 $\dot{Q}$ 表示单位时间内控制体和周围环境以辐射、传导或对流的形式交换的能量。

功传输率 $\dot{W}$ 也称作为功率，环境对控制体做功为正，反之为负。功率的一些重要形式包括：净输入的轴功率 $\dot{W}_{\text{shaft net in}}$、正应力引起的功率 $\dot{W}_{\text{ns}}$ 和切应力引起的功率 $\dot{W}_{\text{ts}}$。通常，切应力引起的功率为零或非常小，以至于可以忽略。

The power transfer associated with normal stresses acting on a fluid particle, $\delta \dot{W}_{ns}$, can be expressed as

作用在流体微团上的正应力引起的功率传输 $\delta \dot{W}_{ns}$ 可表达为

$$\delta \dot{W}_{ns} = \delta \mathbf{F}_{ns} \cdot \mathbf{v} \quad (4.52)$$

where $\delta \mathbf{F}_{ns}$ is the normal stress force, and $\mathbf{v}$ is the fluid particle velocity. For most situations, the fluid normal stress, σ, is simply equal to the negative of fluid pressure, p; that is, $\sigma = -p$. If $\delta \mathbf{F}_{ns}$ is expressed as the product of normal stress, σ, and fluid particle surface area, $\mathbf{n} \cdot \delta A$, the above equation can be written as

式中，$\delta \mathbf{F}_{ns}$ 为正应力产生的力；$\mathbf{v}$ 为流体微团的速度。对于大多数情况，流体正应力等于负的流体压力，即 $\sigma = -p$。如果 $\delta \mathbf{F}_{ns}$ 表示为正应力 σ 和流体微团表面积 $\mathbf{n} \cdot \delta A$ 的乘积，上式可写为

$$\delta \dot{W}_{ns} = \sigma \mathbf{n} \delta A \cdot \mathbf{v} = -p \mathbf{n} \delta A \cdot \mathbf{v} = -p \mathbf{v} \cdot \mathbf{n} \delta A \quad (4.53)$$

The total power transfer due to fluid normal stress can be expressed as

由流体正应力引起的总的功率传输可表达为

$$\dot{W}_{ns} = \int_{cs} \sigma \mathbf{v} \cdot \mathbf{n} dA = \int_{cs} -p \mathbf{v} \cdot \mathbf{n} dA \quad (4.54)$$

According to the above analysis, the first law of thermodynamics for the contents of a control volume [Eq. (4.51)] can be expressed as

根据上述分析，基于控制体的热力学第一定律 [式(4.51)] 可表达为

$$\frac{\partial}{\partial t} \int_{cv} e\rho dV + \int_{cs} e\rho \mathbf{v} \cdot \mathbf{n} dA = \dot{Q}_{net\,in} + \dot{W}_{shaft\,net\,in} + \underbrace{\dot{W}_{ns}}_{=-\int_{cs} p\mathbf{v} \cdot \mathbf{n} dA} + \underbrace{\dot{W}_{ts}}_{=0} \quad (4.55)$$

When Eq. (4.48) is considered with Eq. (4.55), the energy equation can be written as

将式(4.48)代入式(4.55)，变形后得能量方程为

$$\frac{\partial}{\partial t} \int_{cv} e\rho dV + \int_{cs} \left(u + \frac{p}{\rho} + \frac{v^2}{2} + gz\right)\rho \mathbf{v} \cdot \mathbf{n} dA = \dot{Q}_{net\,in} + \dot{W}_{shaft\,net\,in} \quad (4.56)$$

4.4.2 Application of the Energy Equation

4.4.2 能量方程的应用

In Eq. (4.56), the term $\frac{\partial}{\partial t}\int_{cv} e\rho dV$ represents the time rate of change of the total energy stored in the control volume. For steady flow, this term is zero.

式(4.56) 中，$\frac{\partial}{\partial t}\int_{cv} e\rho dV$ 表示储存在控制体内总能量的时间变化率。对于定常流动，这一项为零。

The integrand of $\int_{cs}\left(u + \frac{p}{\rho} + \frac{v^2}{2} + gz\right)\rho \mathbf{v} \cdot \mathbf{n} dA$ can be

只有在流体穿过控制面的地方（$\mathbf{v} \cdot$

nonzero only where fluid crosses the control surface ($\mathbf{v} \cdot \mathbf{n} \neq 0$). If the properties within parentheses, u, p/ρ, $v^2/2$ and gz, are assumed to be uniform over the flow cross-sectional areas, the integration can be written as

$\mathbf{n} \neq 0$), $\int_{cs}\left(u+\dfrac{p}{\rho}+\dfrac{v^2}{2}+gz\right)\rho\mathbf{v}\cdot\mathbf{n}dA$ 中的被积函数不为零。如果括号内的各项（u、p/ρ、$v^2/2$ 和 gz）在流动的横截面上均匀分布，则积分可写成

$$\int_{cs}\left(u+\frac{p}{\rho}+\frac{v^2}{2}+gz\right)\rho\mathbf{v}\cdot\mathbf{n}dA=\sum_{\text{flow out}}\left(u+\frac{p}{\rho}+\frac{v^2}{2}+gz\right)\dot{m}-\sum_{\text{flow in}}\left(u+\frac{p}{\rho}+\frac{v^2}{2}+gz\right)\dot{m} \quad (4.57)$$

For the case with only one inlet and outlet

对于只有一个进出口的情况

$$\int_{cs}\left(u+\frac{p}{\rho}+\frac{v^2}{2}+gz\right)\rho\mathbf{v}\cdot\mathbf{n}dA=\left(u+\frac{p}{\rho}+\frac{v^2}{2}+gz\right)_{\text{out}}\dot{m}_{\text{out}}-\left(u+\frac{p}{\rho}+\frac{v^2}{2}+gz\right)_{\text{in}}\dot{m}_{\text{in}} \quad (4.58)$$

For a steady-in-the-mean flow with only one inlet and outlet, if shaft work is involved, Eq. (4.56) can be simplified with the help of Eq. (4.7) and Eq. (4.58) to form

对于只有一个进出口的时均定常流动，如果包含轴功，根据式(4.7)和式(4.58)，式(4.56)可简化为

$$\dot{m}\left[u_{\text{out}}-u_{\text{in}}+\left(\frac{p}{\rho}\right)_{\text{out}}-\left(\frac{p}{\rho}\right)_{\text{in}}+\frac{v_{\text{out}}^2-v_{\text{in}}^2}{2}+g(z_{\text{out}}-z_{\text{in}})\right]=\dot{Q}_{\text{net in}}+\dot{W}_{\text{shaft net in}} \quad (4.59)$$

Equation (4.59) is the one-dimensional energy equation for steady-in-the-mean flow. This equation is valid for incompressible and compressible flows. With enthalpy ($h=u+p/\rho$), the Eq. (4.59) can be written as

式(4.59)为定常均匀流动的一维能量方程，该方程对不可压缩和可压缩流动均适用。采用焓（$h=u+p/\rho$）的形式，式(4.59)可表达为

$$\dot{m}\left[h_{\text{out}}-h_{\text{in}}+\frac{v_{\text{out}}^2-v_{\text{in}}^2}{2}+g(z_{\text{out}}-z_{\text{in}})\right]=\dot{Q}_{\text{net in}}+\dot{W}_{\text{shaft net in}} \quad (4.60)$$

Equation (4.60) is often used for solving compressible flow problems. Another form of the energy equation that is most often used to solve incompressible flow problems will be introduced in the next section.

式(4.60)通常用于求解可压缩流动问题。下一节将介绍能量方程的另一种形式，它常用于求解不可压缩流动问题。

If the shaft power is zero, the above two equations can be written as

如果轴功率为零，上述两式可写为

$$\dot{m}\left[u_{\text{out}}-u_{\text{in}}+\left(\frac{p}{\rho}\right)_{\text{out}}-\left(\frac{p}{\rho}\right)_{\text{in}}+\frac{v_{\text{out}}^2-v_{\text{in}}^2}{2}+g(z_{\text{out}}-z_{\text{in}})\right]=\dot{Q}_{\text{net in}} \quad (4.61)$$

$$\dot{m}\left[h_{\text{out}}-h_{\text{in}}+\frac{v_{\text{out}}^2-v_{\text{in}}^2}{2}+g(z_{\text{out}}-z_{\text{in}})\right]=\dot{Q}_{\text{net in}} \quad (4.62)$$

【EXAMPLE 4.4】 As shown in Fig. 4.8, steam enters a turbine with a velocity of 27m/s and enthalpy of 3013kJ/kg

【例题 4.4】 如图 4.8 所示，蒸汽以 27m/s 的速度进入透平，其焓

and then leaves the turbine with a velocity of 54m/s and an enthalpy of 2295kJ/kg. If the flow through the turbine is adiabatic and changes in elevation are negligible, determine the work output involved per unit mass of steam through-flow.

值为3013kJ/kg，并以54m/s的速度、2295kJ/kg的焓值离开透平。如果流动过程绝热，且高度变化可以忽略，试确定单位质量蒸汽所输出的功。

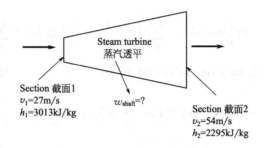

Fig. 4.8　Figure for the example 4.4
图4.8　例题4.4图

SOLUTION　Select a control volume that includes the steam in the turbine from the entrance to the exit as shown in Fig. 4.8. Applying Eq. (4.60) to this control volume, it can be obtained that

解　选择从进口到出口的整个透平作为控制体，如图4.8所示。将式(4.60)应用于控制体，可得

$$\dot{m}\left[h_{out}-h_{in}+\frac{v_{out}^2-v_{in}^2}{2}+\underbrace{g(z_{out}-z_{in})}_{=0(\text{elevation change is negligible})}\right]=\underbrace{\dot{Q}_{net\,in}}_{=0\,(\text{adiabatic flow})}+\dot{W}_{shaft\,net\,in} \quad (4.63)$$

Dividing Eq. (4.63) by the mass flow rate, $\dot{m}$, it can be obtained that

式(4.63)两边同时除以质量流量 $\dot{m}$，得

$$w_{shaft\,net\,in}=\frac{\dot{W}_{shaft\,net\,in}}{\dot{m}}=h_2-h_1+\frac{v_2^2-v_1^2}{2} \quad (4.64)$$

Since $w_{shaft\,net\,out}=-w_{shaft\,net\,in}$, it can be obtained that

由于 $w_{shaft\,net\,out}=-w_{shaft\,net\,in}$，故

$$w_{shaft\,net\,out}=h_1-h_2+\frac{v_1^2-v_2^2}{2}$$

$$w_{shaft\,net\,out}=3013-2295+\frac{(27)^2-(54)^2}{2\times 1000}=717\ (\text{kJ/kg})$$

4.4.3　The Bernoulli Equation

4.4.3　伯努利方程

For incompressible flow, Eq. (4.61) can be written as

对于不可压缩流动，式(4.61)可写为

$$\dot{m}\left[u_{out}-u_{in}+\frac{p_{out}}{\rho}-\frac{p_{in}}{\rho}+\frac{v_{out}^2-v_{in}^2}{2}+g(z_{out}-z_{in})\right]=\dot{Q}_{net\ in} \qquad (4.65)$$

Dividing Eq. (4.65) by the mass flow rate, $\dot{m}$, and rearranging terms, it can be expressed as

式(4.65)两边同时除以质量流量 $\dot{m}$，整理后上式可表达为

$$\frac{p_{out}}{\rho}+\frac{v_{out}^2}{2}+gz_{out}=\frac{p_{in}}{\rho}+\frac{v_{in}^2}{2}+gz_{in}-(u_{out}-u_{in}-q_{net\ in}) \qquad (4.66)$$

where $q_{net\ in}=\dfrac{\dot{Q}_{net\ in}}{\dot{m}}$ is the heat transfer rate per mass flow rate, or heat transfer per unit mass.

式中，$q_{net\ in}=\dfrac{\dot{Q}_{net\ in}}{\dot{m}}$ 表示单位质量流量的传热速率，或单位质量的传热量。

If the viscous effect can be negligible and there is no heat transfer, $u_{out}-u_{in}-q_{net\ in}=0$. The Eq. (4.66) becomes

如果黏性作用可以忽略，且无热量传递，则 $u_{out}-u_{in}-q_{net\ in}=0$。式(4.66)变为

$$\frac{p_{out}}{\rho}+\frac{v_{out}^2}{2}+gz_{out}=\frac{p_{in}}{\rho}+\frac{v_{in}^2}{2}+gz_{in} \qquad (4.67)$$

Eq. (4.67) is called the Bernoulli equation, which is restricted to the following conditions:
- inviscid flow
- steady flow
- incompressible flow
- no heat transfer
- shaft power is zero

式(4.67)称作为伯努利方程。该方程应该满足以下条件：
- 无黏流动
- 定常流动
- 不可压缩流动
- 无热量传递
- 轴功率为零

From comparison of Eq. (4.66) and Eq. (4.67), it can be found that $u_{out}-u_{in}-q_{net\ in}$ represents the loss of energy. Thus, Eq. (4.66) can also be expressed as

比较式(4.66)和式(4.67)可以发现，$u_{out}-u_{in}-q_{net\ in}$ 表示能量损失。因此，式(4.66)也可表达为

$$\frac{p_{out}}{\rho}+\frac{v_{out}^2}{2}+gz_{out}=\frac{p_{in}}{\rho}+\frac{v_{in}^2}{2}+gz_{in}-C_{loss} \qquad (4.68)$$

If the shaft work needs to be considered in the problem, Eq. (4.68) becomes

如果该问题需要考虑轴功，式(4.68)变为

$$\frac{p_{out}}{\rho}+\frac{v_{out}^2}{2}+gz_{out}=\frac{p_{in}}{\rho}+\frac{v_{in}^2}{2}+gz_{in}+w_{shaft\ net\ in}-C_{loss} \qquad (4.69)$$

This equation is called the extended Bernoulli equation. Note that Eq. (4.69) involves energy per unit mass.

该式称为扩展的伯努利方程。注意：式中的能量为单位质量所具有的能量。

If Eq. (4.69) is multiplied by fluid density, ρ, it becomes

式(4.69)两边同时乘以密度 ρ，可得

$$p_{out} + \frac{\rho v_{out}^2}{2} + \gamma z_{out} = p_{in} + \frac{\rho v_{in}^2}{2} + \gamma z_{in} + \rho w_{shaft\ net\ in} - \rho C_{loss} \quad (4.70)$$

where $\gamma = \rho g$ is the specific weight of the fluid. Equation. (4.70) involves energy per unit volume and the units involved are identical with those used for pressure.

式中，$\gamma = \rho g$ 表示流体的重度。式(4.70)中的能量为单位体积具有的能量，其单位与压力单位一致。

If Eq. (4.69) is divided by the acceleration of gravity, g, it can be obtained that

式(4.69)两边同时除以重力加速度 g，可得

$$\frac{p_{out}}{\gamma} + \frac{v_{out}^2}{2g} + z_{out} = \frac{p_{in}}{\gamma} + \frac{v_{in}^2}{2g} + z_{in} + h_s - h_L \quad (4.71)$$

where

其中

$$h_s = w_{shaft\ net\ in}/g = \frac{\dot{W}_{shaft\ net\ in}}{\dot{m}g} = \frac{\dot{W}_{shaft\ net\ in}}{\gamma Q} \text{ and } h_L = C_{loss}/g$$

Equation(4.71) involves energy per unit weight and the units involved are identical with those used for length.

式(4.71)中的能量为单位重量具有的能量，其单位与长度单位一致。

【EXAMPLE 4.5】 An axial-flow ventilating fan shown in Fig. 4.9 is driven by a motor, which delivers 0.35kW of power to the fan blades. The fan produces a axial flow with a diameter of 0.65m and the speed of the flow is 11m/s. The speed of the air in the upstream of the fan can be negligible. Determine the useful work added to the air and the efficiency of this fan.

【例题 4.5】 一个轴流通风机由电机驱动，如图 4.9 所示，电机传输给风机叶片的功率为 0.35kW。风机产生一个直径为 0.65m 的轴向气流，流速为 11m/s。风机上游的空气流速可以忽略不计。试确定传输给空气的有用功及风机的效率。

SOLUTION Select a control volume as is illustrated in Fig. 4.9. The application of Eq. (4.69) to this problem results in

解 选择如图 4.9 所示的控制体，将式(4.69)应用于该问题，得

$$w_{shaft\ net\ in} - C_{loss} = \left(\frac{p_2}{\rho} + \frac{v_2^2}{2} + gz_2\right) - \left(\frac{p_1}{\rho} + \frac{v_1^2}{2} + gz_1\right)$$

$$= \underbrace{\frac{p_2}{\rho} - \frac{p_1}{\rho}}_{\substack{=0 \\ \text{(atmospheric pressures cancel)}}} + \underbrace{gz_2 - gz_1}_{\substack{=0 \\ \text{(no elevation change)}}} + \frac{v_2^2}{2} - \underbrace{\frac{v_1^2}{2}}_{\substack{=0 \\ (v_1 \approx 0)}} \quad (4.72)$$

where $w_{shaft\ net\ in} - C_{loss}$ is the amount of work added to

式中，$w_{shaft\ net\ in} - C_{loss}$ 为输送给

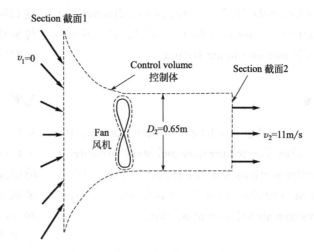

Fig. 4.9 Figure for the example 4.5
图 4.9 例题 4.5 图

the air that produces a useful effect. Equation (4.72) can be written as

空气的有用功。式(4.72) 可写为

$$w_{\text{shaft net in}} - C_{\text{loss}} = \frac{v_2^2}{2} = \frac{11^2}{2} = 60.5 \ (\text{N} \cdot \text{m/kg}) \quad (4.73)$$

The efficiency of this fan is the ratio of amount of work that produces a useful effect to the amount of work delivered to the fan blades. That is

风机的效率等于输送给空气的有用功与输送给风机叶片的功之比，即

$$\eta = \frac{w_{\text{shaft net in}} - C_{\text{loss}}}{w_{\text{shaft net in}}} \quad (4.74)$$

$w_{\text{shaft net in}}$ can be calculated by following equation

$w_{\text{shaft net in}}$ 可通过下式计算

$$w_{\text{shaft net in}} = \frac{\dot{W}_{\text{shaft net in}}}{\dot{m}} \quad (4.75)$$

where $\dot{W}_{\text{shaft net in}}$ is the power delivered to the blades, and $\dot{m}$ is the mass flow rate. For standard air, the fluid density, ρ, is 1.23 kg/m³, and thus

式中，$\dot{W}_{\text{shaft net in}}$ 为输送给叶片的功率；$\dot{m}$ 为质量流量。标准条件下的空气密度为 1.23 kg/m³，因此

$$w_{\text{shaft net in}} = \frac{\dot{W}_{\text{shaft net in}}}{(\rho \pi D_2^2/4)v_2} = \frac{0.35 \times 10^3}{(1.23 \times \pi \times 0.65^2/4) \times 11} = 78.0 \ (\text{N} \cdot \text{m/kg}) \quad (4.76)$$

From Eq. (4.73), Eq. (4.74) and Eq. (4.76), it can be obtained that

由式(4.73)、式(4.74) 和式(4.76) 得

$$\eta = \frac{60.5}{78.0} = 0.776$$

This means that only 77.6% of the power delivered to the air results in a useful effect, and 22.4% of the shaft power is lost due to air friction.

Exercises

4.1 Air flows steadily in a long, straight pipe with the diameter of 0.1m. The temperature and static pressure at inlet and outlet sections are illustrated in Fig. 4.10. If the average air velocity at section 1 is 205m/s, determine the average air velocity at section 2.

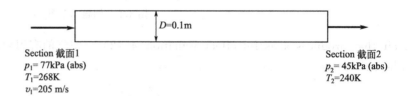

Fig. 4.10 Figure for the exercise 4.1

图 4.10 习题 4.1 图

4.2 The cross-sectional area of the test section of a large water tunnel is $10m^2$. If the test velocity is 15m/s, what volume flow rate capacity in m^3/h is needed?

4.3 Water enters a tank through section 1 and section 2 at rates of 1 and $0.4m^3/min$, respectively (see Fig. 4.11). If the level of the water in the tank remains constant, calculate the average velocity of the flow leaving the tank through section 3.

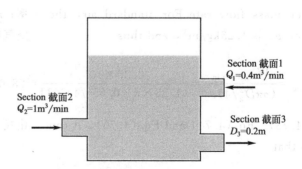

Fig. 4.11 Figure for the exercise 4.3

图 4.11 习题 4.3 图

4.4 It takes you 1min to fill your car's fuel tank with 50L of gasoline. What is the approximate average velocity of the gasoline leaving the nozzle?

4.4 如果给汽车加 50L 的汽油需要 1 分钟的时间，求：汽油流出油枪喷嘴的平均流速大约为多大？

4.5 As shown in Fig. 4.12, a nozzle discharges water into the atmosphere. When the flow rate is $0.1 \text{m}^3/\text{s}$, the gage pressure at the inlet is 40kPa. The nozzle has a weight of 200N, and the volume of water in the nozzle is 0.012m^3. Determine the anchoring force required to hold the nozzle.

4.5 如图 4.12 所示，喷嘴向空中喷出水。当流量为 $0.1\text{m}^3/\text{s}$ 时，进口的表压为 40kPa。喷嘴重 200N，喷嘴内水的体积为 0.012m^3。试确定握住喷嘴所需的力。

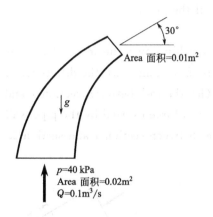

Fig. 4.12 Figure for the exercise 4.5
图 4.12 习题 4.5 图

4.6 Water flows through a tee shown in Fig. 4.13. The exit speed is 15m/s. If viscous effects and gravity are negligible, determine the x and y components of the force that the pipe exerts on the tee.

4.6 水流经如图 4.13 所示的三通，出流速度为 15m/s。忽略黏性作用和重力，试确定圆管作用在三通上的 x 和 y 方向上的力。

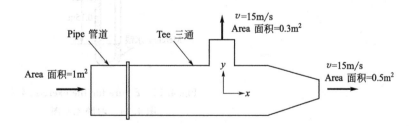

Fig. 4.13 Figure for the exercise 4.6
图 4.13 习题 4.6 图

4.7 A truck carrying chickens needs to cross a bridge. The empty truck is within the weight limits; with the chickens it is overweight. It is suggested that if one

4.7 装有鸡的卡车需要从桥上通过，空车在桥的荷载范围内，装上鸡就超载。有人建议如果让所有

could get the chickens to fly in the truck it would be safe to cross the bridge. Do you agree?

的鸡在卡车内飞起来，卡车就可以安全从桥上通过。你认为对吗？

4.8 Water enters a pump impeller radially with the flow rate of Q. It leaves the impeller with a tangential component of absolute velocity of v_t. The impeller exit diameter is D_2, and the impeller speed is ω. If the total pressure rise across the impeller is p, determine the loss of available energy across the impeller and the hydraulic efficiency of the pump.

4.8 水以流量 Q 沿径向进入泵叶轮，离开叶轮时的绝对速度的切向分量为 v_t，叶轮出口直径为 D_2，叶轮转速为 ω。如果通过叶轮的总压升高 p，试确定叶轮内的能量损失以及泵的水力效率。

4.9 Water flows steadily down the inclined pipe as indicated in Fig. 4.14. Determine (a) the difference in pressure p_1-p_2, (b) the loss between section 1 and section 2, (c) the axial force exerted by the pipe wall on the flowing water between section 1 and section 2.

4.9 水沿倾斜管向下流动，如图 4.14 所示。试确定：(a) 压差 p_1-p_2，(b) 截面 1 和截面 2 之间的能量损失，(c) 截面 1 和截面 2 之间管壁作用在流体上的轴向力。

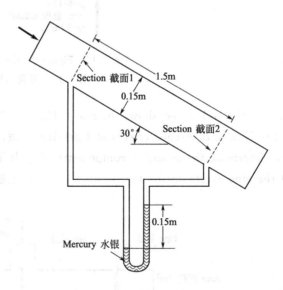

Fig. 4.14 Figure for the exercise 4.9
图 4.14 习题 4.9 图

4.10 Oil ($SG=0.86$) flows in an inclined pipe at a rate of $0.14 \text{m}^3/\text{s}$ as shown in Fig. 4.15. If the differential reading in the mercury manometer is 0.9m, calculate the power that the pump supplies to the oil (the head losses are negligible).

4.10 比重为 0.86 的油以 $0.14\text{m}^3/\text{s}$ 的流量在倾斜管内流动，如图 4.15 所示。如果水银压力计内的压差读数为 0.9m，试计算泵输送给油的功率（忽略水头损失）。

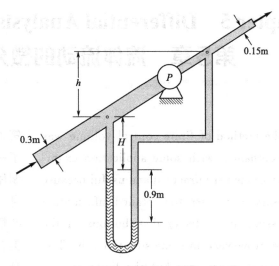

Fig. 4.15　Figure for the exercise 4.10
图 4.15　习题 4.10 图

Chapter 5　Differential Analysis of Fluid Flow
第5章　流体流动的微分分析

In chapter 4 the method of finite control volume was introduced accompanied with some application examples. This approach is very practical and useful because it is not necessary to concern the detailed information about the pressure and velocity distributions in the control volume. However, in some situations the flow details are of great importance but the finite control volume method can not yield the desired information. So some flow equations are desired to be established for a point or a very small region (infinitesimal control volume) in a given flow field. To distinguish it from the finite control volume, this approach is commonly called differential analysis because the governing equations are differential equations. The derivation of the differential equations and some applications will be introduced in this chapter.

第4章介绍了有限控制体法并给出了一些实例，由于不需要关注控制体内压力和速度分布的详细信息，该方法非常实用。然而，有些情况下流动的细节非常重要，但有限控制体法无法得到所需信息。此时，需要针对流场中的点或非常小的区域（微控制体）来建立流动方程。因为所建立的控制方程为微分方程，为了区别于有限控制体法，该方法通常称为微分分析法。本章将介绍微分方程的推导及其应用。

5.1　Conservation of Mass

5.1　质量守恒

As introduced in section 4.1.1, the conservation of mass for control volume can be expressed as

如4.1.1节所介绍的，基于控制体的质量守恒可表达为

$$\frac{\partial}{\partial t}\int_{cv}\rho dV + \int_{cs}\rho \mathbf{v} \cdot \mathbf{n} dA = 0 \quad (5.1)$$

This equation is known as the continuity equation. The first integral on the left side of Eq. (5.1) represents the rate of mass change in the control volume, and the second integral represents the net mass flow rate through the control surface. To obtain the continuity equation in the differential form, Eq. (5.1) is applied to an infinitesimal control volume.

该方程通常称为连续性方程。式(5.1)左边的第一个积分项表示控制体内质量的变化率，第二个积分项表示通过控制面的净质量流量。为了得到微分形式的连续性方程，将式(5.1)应用于微控制体。

5.1.1 Continuity Equation in Differential Form

The control volume can be taken as the small, stationary cubical element shown in Fig. 5.1(a). At the center of the element the fluid density is ρ and the velocity has components v_x, v_y and v_z. Since the element is small, the volume integral in Eq. (5.1) can be expressed as

5.1.1 微分形式的连续性方程

控制体可视为微小的静止立方体单元，如图5.1(a)所示。微元中心的密度为 ρ，速度分量为 v_x、v_y 和 v_z。由于单元非常小，式(5.1)中的体积分可以表达为

$$\frac{\partial}{\partial t}\int_{cv}\rho dV \approx \frac{\partial \rho}{\partial t}\delta x \delta y \delta z \qquad (5.2)$$

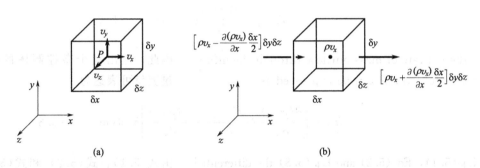

Fig. 5.1 An infinitesimal control volume for the development of continuity equation

图 5.1 推导连续性方程用的微元

The rate of mass flow through the surfaces of the element can be obtained by considering the flow in each of the coordinate directions separately. The flow in the x direction is shown in Fig. 5.1(b). ρv_x represents the x component of the rate of mass flow per unit area at the center of the element, then on the right surface

分别考虑各个坐标轴方向上的流动，可得到通过微元表面的质量流量。图5.1(b)显示了 x 方向的流动。ρv_x 表示 x 方向上微元中心单位面积上的质量流量，微元右侧单位面积上的质量流量为

$$\rho v_x|_{x+(\delta x/2)} = \rho v_x + \frac{\partial(\rho v_x)}{\partial x}\frac{\delta x}{2} \qquad (5.3)$$

and on the left surface

左侧单位面积上的质量流量为

$$\rho v_x|_{x-(\delta x/2)} = \rho v_x - \frac{\partial(\rho v_x)}{\partial x}\frac{\delta x}{2} \qquad (5.4)$$

A Taylor series expansion of ρv_x is used in the above two equations and the higher order terms are neglected. Multiplied by the area $\delta y \delta z$ in the right-hand sides of Eq. (5.3) and Eq. (5.4), the mass transfer

上面两式采用了泰勒级数展开式，并忽略了高阶项。式(5.3)和式(5.4)的等号右边乘以面积 $\delta y \delta z$，得到通过左右两侧面的质量流量，

rate through the right and left surfaces of the element are obtained, as illustrated in Fig. 5.1(b). Combination of Eq. (5.3) and Eq. (5.4) gives the net rate of mass outflow in x direction

如图 5.1(b) 所示。结合式(5.3) 和式(5.4)，x 方向上净流出质量流量为

$$\dot{m}_{\text{out}x} = \left[\rho v_x + \frac{\partial(\rho v_x)}{\partial x}\frac{\delta x}{2}\right]\delta y \delta z - \left[\rho v_x - \frac{\partial(\rho v_x)}{\partial x}\frac{\delta x}{2}\right]\delta y \delta z = \frac{\partial(\rho v_x)}{\partial x}\delta x \delta y \delta z \quad (5.5)$$

Similarly, the net rates of mass outflow in the y and z directions can be expressed as

类似地，y、z 方向上净流出质量流量可表达为

$$\dot{m}_{\text{out}y} = \frac{\partial(\rho v_y)}{\partial y}\delta x \delta y \delta z \quad (5.6)$$

$$\dot{m}_{\text{out}z} = \frac{\partial(\rho v_z)}{\partial z}\delta x \delta y \delta z \quad (5.7)$$

Thus, the net rate of mass outflow through the infinitesimal control volume can be expressed as

因此，流出整个微控制体的净质量流量可表达为

$$\dot{m}_{\text{cv}} = \left[\frac{\partial(\rho v_x)}{\partial x} + \frac{\partial(\rho v_y)}{\partial y} + \frac{\partial(\rho v_z)}{\partial z}\right]\delta x \delta y \delta z \quad (5.8)$$

From Eq. (5.1), Eq. (5.2) and Eq. (5.8) the differential equation for conservation of mass can be expressed as

由式(5.1)、式(5.2) 和式(5.8)，质量守恒的微分方程可表达为

$$\frac{\partial \rho}{\partial t} + \frac{\partial(\rho v_x)}{\partial x} + \frac{\partial(\rho v_y)}{\partial y} + \frac{\partial(\rho v_z)}{\partial z} = 0 \quad (5.9)$$

The continuity equation is one of the fundamental equations of fluid mechanics. It is valid for both steady and unsteady flows, and for compressible and incompressible fluids. In vector notation, Eq. (5.9) can be written as

连续性方程是流体力学中的基本方程之一，它对定常、非定常流动，可压缩、不可压缩流体，均适用。采用矢量形式，式(5.9) 可写成

$$\frac{\partial \rho}{\partial t} + \nabla \cdot \rho \mathbf{v} = 0 \quad (5.10)$$

For steady flow of compressible fluids

对于可压缩流体的定常流动

$$\nabla \cdot \rho \mathbf{v} = 0$$

or

或

$$\frac{\partial(\rho v_x)}{\partial x} + \frac{\partial(\rho v_y)}{\partial y} + \frac{\partial(\rho v_z)}{\partial z} = 0 \quad (5.11)$$

For incompressible fluids the fluid density is a constant so that Eq. (5.10) becomes

对于不可压缩流体，密度为常数，式(5.10) 变为

$$\nabla \cdot \mathbf{v} = 0$$

or 或

$$\frac{\partial v_x}{\partial x} + \frac{\partial v_y}{\partial y} + \frac{\partial v_z}{\partial z} = 0 \quad (5.12)$$

Equation (5.12) is appropriate for both steady and unsteady flow of incompressible fluids. It means the volumetric dilatation rate equal to zero.

式(5.12)适用于不可压缩流体的定常和非定常流动,它表示体积膨胀率为零。

【EXAMPLE 5.1】 There is an incompressible and steady flow field with the three velocity components as

【例题 5.1】 不可压缩定常流场的三个速度分量为

$$v_x = xy + yz + zx$$

$$v_y = x^2 + y^2 + z^2$$

$$v_z = ?$$

Determine the velocity component, v_z, in z direction to satisfy the requirement of the continuity equation.

试确定 z 方向速度分量 v_z 以满足连续性方程。

SOLUTION For an incompressible fluid, the velocity distribution must satisfy the following continuity equation

解 对于不可压缩流体,速度分布必须满足如下连续性方程

$$\frac{\partial v_x}{\partial x} + \frac{\partial v_y}{\partial y} + \frac{\partial v_z}{\partial z} = 0$$

For the given velocity components

对已给定的速度分量

$$\frac{\partial v_x}{\partial x} = y + z \quad \text{and} \quad \frac{\partial v_y}{\partial y} = 2y$$

so that the required expression for $\partial v_z / \partial z$ is

因此,所求的关于 $\partial v_z / \partial z$ 的表达式为

$$\frac{\partial v_z}{\partial z} = -2y - (y + z) = -3y - z$$

Integration with respect to z yields

上式对 z 积分,得

$$v_z = -3yz - \frac{z^2}{2} + f(x, y)$$

The velocity component v_z with any form of the function $f(y, z)$ can satisfy the continuity equation, so the specific form of v_z can not be determined according to the given information.

任何形式的 $f(y, z)$ 均可使 v_z 满足连续性方程,所以根据已知信息无法确定 z 方向的速度分量。

5.1.2 Continuity Equation in Cylindrical Coordinates

In cylindrical coordinates, the velocity components are the radial velocity, v_r, the tangential velocity, v_θ, and the axial velocity, v_z, as sketched in Fig. 5.2. Thus, the velocity at any point P can be expressed as

$$\mathbf{v} = v_r \mathbf{e}_r + v_\theta \mathbf{e}_\theta + v_z \mathbf{e}_z \quad (5.13)$$

where $\mathbf{e}_r$, $\mathbf{e}_\theta$ and $\mathbf{e}_z$ are the unit vectors in the r, θ and z directions, respectively.

5.1.2 柱坐标系中的连续性方程

柱坐标系中的各速度分量（径向速度 v_r、切向速度 v_θ、轴向速度 v_z）如图 5.2 所示，任意点 P 的速度可表达为

式中，$\mathbf{e}_r$、$\mathbf{e}_\theta$ 和 $\mathbf{e}_z$ 分别为 r、θ 和 z 方向的单位矢量。

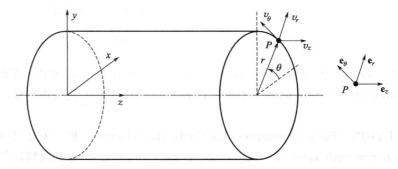

Fig. 5.2 The representation of velocity components in cylindrical coordinates

图 5.2 柱坐标系中速度分量的表示

The differential form of the continuity equation in cylindrical coordinates is

柱坐标系中微分形式的连续性方程为

$$\frac{\partial \rho}{\partial t} + \frac{1}{r}\frac{\partial (r\rho v_r)}{\partial r} + \frac{1}{r}\frac{\partial (\rho v_\theta)}{\partial \theta} + \frac{\partial (\rho v_z)}{\partial z} = 0 \quad (5.14)$$

For steady compressible flow

对于定常可压缩流动

$$\frac{1}{r}\frac{\partial (r\rho v_r)}{\partial r} + \frac{1}{r}\frac{\partial (\rho v_\theta)}{\partial \theta} + \frac{\partial (\rho v_z)}{\partial z} = 0 \quad (5.15)$$

For incompressible fluids (for steady or unsteady flow)

对于不可压缩流体（定常或非定常）

$$\frac{1}{r}\frac{\partial (rv_r)}{\partial r} + \frac{1}{r}\frac{\partial v_\theta}{\partial \theta} + \frac{\partial v_z}{\partial z} = 0 \quad (5.16)$$

5.2 Conservation of Momentum

The momentum equation for the system is

5.2 动量守恒

系统的动量方程为

$$\mathbf{F} = \left.\frac{D\mathbf{P}}{Dt}\right|_{sys} \quad (5.17)$$

where **F** is the resultant force acting on a system, and **P** is the momentum expressed as

$$\mathbf{P} = \int_{sys} \mathbf{v}\,dm$$

and the operator $D(\ )/Dt$ is the material derivative. As introduced in section 4.2.1, the momentum equation based on the control volume is

$$\sum \mathbf{F}_{cv} = \frac{\partial}{\partial t}\int_{cv} \mathbf{v}\rho\,dV + \int_{cs} \mathbf{v}\rho\mathbf{v}\cdot\mathbf{n}\,dA \quad (5.18)$$

To obtain the differential form of the momentum equation, we can either apply Eq. (5.17) to a differential system consisting of a mass, δm, or apply Eq. (5.18) to an infinitesimal control volume, δV. Since it is simpler to use the system approach, the application of Eq. (5.17) to the differential system yields

$$\delta \mathbf{F} = \frac{D(\mathbf{v}\delta m)}{Dt}$$

where $\delta \mathbf{F}$ is the resultant force acting on δm. Using this system approach δm can be regarded as a constant, so

$$\delta \mathbf{F} = \delta m \frac{D\mathbf{v}}{Dt}$$

where $D\mathbf{v}/Dt$ is the acceleration, **a**, of the element. Thus

$$\delta \mathbf{F} = \delta m\, \mathbf{a} \quad (5.19)$$

which is simply Newton's second law applied to the mass δm.

式中，**F** 为作用在系统上的合力，**P** 是系统的动量，表达为

$D(\ )/Dt$ 为物质导数。如 4.2.1 节所介绍的，基于控制体的动量方程为

为了得到微分形式的动量方程，可以将式（5.17）应用于质量为 δm 的微系统，也可将式（5.18）应用于体积为 δV 的微控制体。由于采用系统的方法更为方便，将式（5.17）应用于微系统，有

式中，$\delta \mathbf{F}$ 为作用在 δm 上的合力。采用系统的方法，δm 可以看作常数，所以有

式中，$D\mathbf{v}/Dt$ 为微元的加速度 **a**。因此

上式即为应用于质量 δm 的牛顿第二定律。

5.2.1 Forces Acting on the Differential Element

In general, the forces that exert on the differential element can be classified as surface forces and body forces, and the former act on the surface of the element while the

5.2.1 作用在微元上的力

通常，作用在微元上的力可分为面力和体力，面力作用在微元的表面，体力分布在整个微元上。

later distribute throughout the element. The body force, $\delta \mathbf{F}_b$, may be induced by the gravitational field, magnetic field or electric field. If only the gravity is considered, $\delta \mathbf{F}_b$ can be expressed as

$$\delta \mathbf{F}_b = \delta m \mathbf{g} \qquad (5.20)$$

where $\mathbf{g}$ is the vector representation of the acceleration of gravity. In component form, $\delta \mathbf{F}_b$ can be expressed as

$$\delta F_{bx} = \delta m g_x \qquad (5.21a)$$
$$\delta F_{by} = \delta m g_y \qquad (5.21b)$$
$$\delta F_{bz} = \delta m g_z \qquad (5.21c)$$

where g_x, g_y and g_z are the components of $\mathbf{g}$ in the x, y and z directions, respectively.

As shown in Fig. 5.3, the surface force, $\delta \mathbf{F}_s$, acting on a small area, δA, can be resolved into three components, δF_n, δF_1 and δF_2, where δF_n is normal to the area, δA, δF_1 and δF_2 are parallel to the area and orthogonal to each other. The normal stress, σ_n, is defined as

$$\sigma_n = \lim_{\delta A \to 0} \frac{\delta F_n}{\delta A}$$

and the shearing stresses are defined as

$$\tau_1 = \lim_{\delta A \to 0} \frac{\delta F_1}{\delta A}, \tau_2 = \lim_{\delta A \to 0} \frac{\delta F_2}{\delta A}$$

体力 $\delta \mathbf{F}_b$ 可由重力场、磁场或电场等引起。若只考虑重力，则 $\delta \mathbf{F}_b$ 可表达为

式中，$\mathbf{g}$ 为重力加速度的矢量。采用分量的形式，$\delta \mathbf{F}_b$ 可表达为

式中，g_x、g_y 和 g_z 分别为 $\mathbf{g}$ 在 x、y 和 z 三个方向上的分量。

如图 5.3 所示，作用在微元面 δA 上的面力 $\delta \mathbf{F}_s$ 可以分解为三个分量：δF_n、δF_1 和 δF_2。其中，δF_n 垂直于面 δA；δF_1 和 δF_2 平行于面 δA，且相互垂直。正应力 σ_n 定义为

切应力定义为

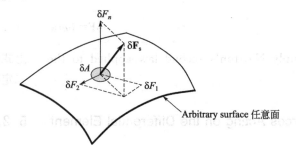

Fig. 5.3 Components of force acting on an arbitrary differential area
图 5.3 任意微元面上的力的分量

Once the orientation of the area is specified, the force per unit area at a point can be characterized by a normal stress and two shearing stresses. For convenience, the stress acting planes are usually considered to be parallel to the coordinate planes, as shown in Fig. 5.4. A double subscript notation is used to identify the stress components on different surfaces. The first subscript indicates the normal direction of the surface on which the stress acts, and the second subscript indicates the direction of the stress. Thus, normal stresses have the same two subscripts, while the two subscripts for the shearing stresses are different.

一旦指定了面的方向，某一点单位面积上的力就可以用一个正应力和两个切应力来表征。为便于分析，认为应力作用面平行于坐标平面，如图5.4所示。采用双下标符号来辨识不同面上的应力分量，第一个下标表示应力作用面的法向，第二个下标表示应力的作用方向。因此，正应力的两个下标相同，切应力的两个下标不同。

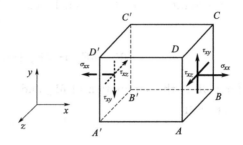

Fig. 5.4 Double subscript notation for stresses
图5.4 应力的双下标符号

The sign for the stresses is defined as follows: If the outward normal of the surface points in the positive coordinate direction, the stresses directing in the positive coordinate directions are considered positive; on the contrary, they are negative. If the outward normal of the surface points in the negative coordinate direction, the stresses directing in the negative coordinate directions are considered positive; on the contrary, they are negative. Thus, the stresses shown in Fig. 5.4 are considered to be positive.

应力正负号的规定如下：若应力所在平面的外法线方向与坐标轴的正向一致，则指向坐标轴正向的应力为正，反之为负；若应力所在平面的外法线方向与坐标轴的正向相反，则指向坐标轴负向的应力为正，反之为负。因此，图5.4中所示的应力均为正。

In general the stresses may vary with the position. Thus, the stresses on the six faces can be expressed using the stresses at the center of the element and their gradients in the coordinate directions, as shown in Fig. 5.5. The surface forces can be expressed in terms of the stresses acting on the faces of the element. The resultant force in one direction can be obtained by sum-

通常应力会随位置而变化，因此，根据微元体中心的应力及其在各坐标方向上的梯度，可以写出六个面上的应力，如图5.5所示。根据作用在微元各面上的应力可以写出各方向上的面力。某一个方向上的各个力相加就可以得到该

ming all the forces in this direction. For example, the resultant surface force in the x direction is

方向上的合力。例如，x 方向上面力的合力为

$$\delta F_{sx} = \left(\frac{\partial \sigma_{xx}}{\partial x} + \frac{\partial \tau_{yx}}{\partial y} + \frac{\partial \tau_{zx}}{\partial z}\right)\delta x \delta y \delta z \quad (5.22a)$$

Similarly, the resultant surface forces in the y and z directions can be expressed as

类似地，y 和 z 方向上面力的合力为

$$\delta F_{sy} = \left(\frac{\partial \tau_{xy}}{\partial x} + \frac{\partial \sigma_{yy}}{\partial y} + \frac{\partial \tau_{zy}}{\partial z}\right)\delta x \delta y \delta z \quad (5.22b)$$

$$\delta F_{sz} = \left(\frac{\partial \tau_{xz}}{\partial x} + \frac{\partial \tau_{yz}}{\partial y} + \frac{\partial \sigma_{zz}}{\partial z}\right)\delta x \delta y \delta z \quad (5.22c)$$

In vector form the resultant surface force can be expressed as

采用矢量的形式，面力的合力可以表达为

$$\delta \mathbf{F}_s = \delta F_{sx}\mathbf{i} + \delta F_{sy}\mathbf{j} + \delta F_{sz}\mathbf{k} \quad (5.23)$$

So the resultant force $\delta \mathbf{F}$ is the summation of $\delta \mathbf{F}_b$ and $\delta \mathbf{F}_s$, that is, $\delta \mathbf{F} = \delta \mathbf{F}_s + \delta \mathbf{F}_b$.

合力 $\delta \mathbf{F}$ 为面力 $\delta \mathbf{F}_s$ 和体力 $\delta \mathbf{F}_b$ 之和，即 $\delta \mathbf{F} = \delta \mathbf{F}_s + \delta \mathbf{F}_b$。

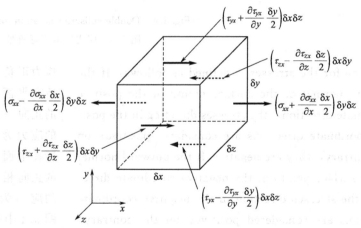

Fig. 5.5 Surface forces in the x direction acting on a fluid element
图 5.5 作用在流体微元上的 x 方向的面力

5.2.2 Equations of Motion

5.2.2 运动方程

Equation (5.21) and Eq. (5.22) can be used in conjunction with Eq. (5.19) to develop the equations of motion. In component form Eq. (5.19) can be written as

式(5.21) 和式(5.22) 与式(5.19) 联立，可以得到运动方程。采用分量的形式，式(5.19) 可以写成

$$\delta F_x = \delta m a_x$$

$$\delta F_y = \delta m a_y$$

$$\delta F_z = \delta m a_z$$

where $\delta m = \rho \delta x \delta y \delta z$, and the acceleration components are given by Eq. (3.4). Canceling out the element volume, $\delta x \delta y \delta z$, it now follows that

式中，$\delta m = \rho \delta x \delta y \delta z$，加速度分量由式（3.4）给出。消去微元体积 $\delta x \delta y \delta z$，得

$$\rho g_x + \frac{\partial \sigma_{xx}}{\partial x} + \frac{\partial \tau_{yx}}{\partial y} + \frac{\partial \tau_{zx}}{\partial z} = \rho \left(\frac{\partial v_x}{\partial t} + v_x \frac{\partial v_x}{\partial x} + v_y \frac{\partial v_x}{\partial y} + v_z \frac{\partial v_x}{\partial z} \right) \quad (5.24a)$$

$$\rho g_y + \frac{\partial \tau_{xy}}{\partial x} + \frac{\partial \sigma_{yy}}{\partial y} + \frac{\partial \tau_{zy}}{\partial z} = \rho \left(\frac{\partial v_y}{\partial t} + v_x \frac{\partial v_y}{\partial x} + v_y \frac{\partial v_y}{\partial y} + v_z \frac{\partial v_y}{\partial z} \right) \quad (5.24b)$$

$$\rho g_z + \frac{\partial \tau_{xz}}{\partial x} + \frac{\partial \tau_{yz}}{\partial y} + \frac{\partial \sigma_{zz}}{\partial z} = \rho \left(\frac{\partial v_z}{\partial t} + v_x \frac{\partial v_z}{\partial x} + v_y \frac{\partial v_z}{\partial y} + v_z \frac{\partial v_z}{\partial z} \right) \quad (5.24c)$$

Equation (5.24) is the general differential equation of motion for fluid. However, before using the equation to solve specific problems, some additional information about the stresses must be obtained.

式（5.24）为一般形式的流体运动微分方程。然而，在采用该方程求解具体问题之前，必须得到关于应力的附加信息。

For an inviscid flow in which all the shearing stresses are zero, and the normal stresses are replaced by $-p$, the Eq. (5.24) are simplified to be

对于无黏流动，所有的切应力为零，正应力用 $-p$ 代替。式（5.24）简化为

$$\rho g_x - \frac{\partial p}{\partial x} = \rho \left(\frac{\partial v_x}{\partial t} + v_x \frac{\partial v_x}{\partial x} + v_y \frac{\partial v_x}{\partial y} + v_z \frac{\partial v_x}{\partial z} \right) \quad (5.25a)$$

$$\rho g_y - \frac{\partial p}{\partial y} = \rho \left(\frac{\partial v_y}{\partial t} + v_x \frac{\partial v_y}{\partial x} + v_y \frac{\partial v_y}{\partial y} + v_z \frac{\partial v_y}{\partial z} \right) \quad (5.25b)$$

$$\rho g_z - \frac{\partial p}{\partial z} = \rho \left(\frac{\partial v_z}{\partial t} + v_x \frac{\partial v_z}{\partial x} + v_y \frac{\partial v_z}{\partial y} + v_z \frac{\partial v_z}{\partial z} \right) \quad (5.25c)$$

These equations are commonly referred to as Eulerian equations of motion. In vector form Euler's equations can be expressed as

该方程称为欧拉运动方程。采用矢量的形式，可表达为

$$\rho \mathbf{g} - \nabla p = \rho \left[\frac{\partial \mathbf{v}}{\partial t} + (\mathbf{v} \cdot \nabla) \mathbf{v} \right] \quad (5.26)$$

Since there are nonlinear velocity terms (such as $v_x \partial v_x / \partial x$, $v_y \partial v_x / \partial y$, etc.) in the convective acceleration, the Eulerian equations are nonlinear partial differential equations. Although Eq. (5.25) is much simpler than the

由于对流加速度中存在非线性速度项，如：$v_x \partial v_x / \partial x$、$v_y \partial v_x / \partial y$ 等，所以欧拉方程是一组非线性偏微分方程。尽管式（5.25）比

general equations of motion [Eq. (5.24)], it is still impossible to obtain a general analytical solution for the pressure and velocity in an inviscid flow field.

5.3 Viscous Flow

Equation (5.24) involves 9 unknown variants (6 stress components and 3 velocity components), while there are only 4 equations (1 continuity equation and 3 equations of motion), and therefore the equations are not closed. It is necessary to establish a relationship between the stresses and velocities.

5.3.1 Stress-Deformation Relationships

For incompressible Newtonian fluids it is known that the stresses are linearly related to the rates of deformation and can be expressed in orthogonal coordinates as

$$\sigma_{xx} = -p + 2\mu \frac{\partial v_x}{\partial x} \quad (5.27\text{a})$$

$$\sigma_{yy} = -p + 2\mu \frac{\partial v_y}{\partial y} \quad (5.27\text{b})$$

$$\sigma_{zz} = -p + 2\mu \frac{\partial v_z}{\partial z} \quad (5.27\text{c})$$

$$\tau_{xy} = \tau_{yx} = \mu \left(\frac{\partial v_x}{\partial y} + \frac{\partial v_y}{\partial x} \right) \quad (5.27\text{d})$$

$$\tau_{yz} = \tau_{zy} = \mu \left(\frac{\partial v_y}{\partial z} + \frac{\partial v_z}{\partial y} \right) \quad (5.27\text{e})$$

$$\tau_{zx} = \tau_{xz} = \mu \left(\frac{\partial v_z}{\partial x} + \frac{\partial v_x}{\partial z} \right) \quad (5.27\text{f})$$

where p is the pressure, and $-p = (\sigma_{xx} + \sigma_{yy} + \sigma_{zz})/3$. For viscous fluids in motion, the normal stresses are not necessarily the same in different directions, thus, the pressure is defined as the average of the three normal stresses. For fluids at rest or inviscid fluids, the normal stresses are equal in all directions.

一般形式的运动方程［式(5.24)］要简单，但是仍然无法获得无黏流场中的压力和速度的解析解。

5.3 黏性流动

式(5.24)包含9个未知量（6个应力分量和3个速度分量），而只有4个方程（1个连续性方程和3个运动方程），因此方程组不封闭，必须建立应力和速度之间的关系。

5.3.1 应力-变形关系

对于不可压缩牛顿流体，应力和变形速率线性相关，在直角坐标系中可表达为

式中，p为压力，等于负的正应力的平均值。对于运动的黏性流体，不同方向上的正应力未必相等，因此定义压力为三个正应力的平均值；对于静止的流体或无黏流体，各方向上的正应力相等。

In cylindrical coordinates the stresses for incompressible Newtonian fluids can be expressed as

柱坐标系中不可压缩牛顿流体的应力可表示为

$$\sigma_{rr} = -p + 2\mu \frac{\partial v_r}{\partial r} \quad (5.28a)$$

$$\sigma_{\theta\theta} = -p + 2\mu \left(\frac{1}{r} \frac{\partial v_\theta}{\partial \theta} + \frac{v_r}{r} \right) \quad (5.28b)$$

$$\sigma_{zz} = -p + 2\mu \frac{\partial v_z}{\partial z} \quad (5.28c)$$

$$\tau_{r\theta} = \tau_{\theta r} = \mu \left[r \frac{\partial}{\partial r} \left(\frac{v_\theta}{r} \right) + \frac{1}{r} \frac{\partial v_r}{\partial \theta} \right] \quad (5.28d)$$

$$\tau_{\theta z} = \tau_{z\theta} = \mu \left(\frac{\partial v_\theta}{\partial z} + \frac{1}{r} \frac{\partial v_z}{\partial \theta} \right) \quad (5.28e)$$

$$\tau_{zr} = \tau_{rz} = \mu \left(\frac{\partial v_r}{\partial z} + \frac{\partial v_z}{\partial r} \right) \quad (5.28f)$$

The subscript has a meaning similar to that of stresses expressed in orthogonal coordinates.

下标的含义和直角坐标系中的应力下标的含义类似。

5.3.2 The Naiver-Stokes Equations

5.3.2 纳维-斯托克斯方程

The stresses as defined by Eq. (5.27) can be substituted into the differential equations of motion [Eq. (5.24)] and simplified by using the continuity equation [Eq. (5.12)] to obtain

将式(5.27)中的应力代入运动微分方程[式(5.24)]，并采用连续性方程[式(5.12)]进行简化后得

$$\rho \left(\frac{\partial v_x}{\partial t} + v_x \frac{\partial v_x}{\partial x} + v_y \frac{\partial v_x}{\partial y} + v_z \frac{\partial v_x}{\partial z} \right) = -\frac{\partial p}{\partial x} + \rho g_x + \mu \left(\frac{\partial^2 v_x}{\partial x^2} + \frac{\partial^2 v_x}{\partial y^2} + \frac{\partial^2 v_x}{\partial z^2} \right) \quad (5.29a)$$

$$\rho \left(\frac{\partial v_y}{\partial t} + v_x \frac{\partial v_y}{\partial x} + v_y \frac{\partial v_y}{\partial y} + v_z \frac{\partial v_y}{\partial z} \right) = -\frac{\partial p}{\partial y} + \rho g_y + \mu \left(\frac{\partial^2 v_y}{\partial x^2} + \frac{\partial^2 v_y}{\partial y^2} + \frac{\partial^2 v_y}{\partial z^2} \right) \quad (5.29b)$$

$$\rho \left(\frac{\partial v_z}{\partial t} + v_x \frac{\partial v_z}{\partial x} + v_y \frac{\partial v_z}{\partial y} + v_z \frac{\partial v_z}{\partial z} \right) = -\frac{\partial p}{\partial z} + \rho g_z + \mu \left(\frac{\partial^2 v_z}{\partial x^2} + \frac{\partial^2 v_z}{\partial y^2} + \frac{\partial^2 v_z}{\partial z^2} \right) \quad (5.29c)$$

where the left side shows the acceleration terms and the right shows the force terms. These equations are commonly called the Navier-Stokes equations. The three equations of motion [Eq. (5.29)] and the conservation of mass equation [Eq. (5.12)] provide a

上式中，等号左边是加速度项，等号右边是力项。该方程通常称为纳维-斯托克斯方程。三个运动方程[式(5.29)]和连续性方程[式(5.12)]构成了不可压缩牛顿

complete mathematical description of the flow of incompressible Newtonian fluids. There are four unknown variants (v_x, v_y, v_z and p) in these four equations, and therefore the equations are closed. But it is still very difficult to get exact solutions from these nonlinear, second-order, partial differential equations, except in a few instances.

流体流动的完整数学描述。四个方程包含四个未知量（v_x、v_y、v_z 和 p），方程是封闭的。但是，除少数情况外，从这些非线性二阶偏微分方程中，仍很难得到精确的解。

In cylindrical coordinate system, the Navier-Stokes equation can be written as

在柱坐标系中，纳维-斯托克斯方程为

$$\rho\left(\frac{\partial v_r}{\partial t}+v_r\frac{\partial v_r}{\partial r}+\frac{v_\theta}{r}\frac{\partial v_r}{\partial \theta}-\frac{v_\theta^2}{r}+v_z\frac{\partial v_r}{\partial z}\right)=-\frac{\partial p}{\partial r}+\rho g_r+\mu\left[\frac{1}{r}\frac{\partial}{\partial r}\left(r\frac{\partial v_r}{\partial r}\right)-\frac{v_r}{r^2}+\frac{1}{r^2}\frac{\partial^2 v_r}{\partial \theta^2}-\frac{2}{r^2}\frac{\partial v_\theta}{\partial \theta}+\frac{\partial^2 v_r}{\partial z^2}\right] \quad (5.30a)$$

$$\rho\left(\frac{\partial v_\theta}{\partial t}+v_r\frac{\partial v_\theta}{\partial r}+\frac{v_\theta}{r}\frac{\partial v_\theta}{\partial \theta}+\frac{v_r v_\theta}{r}+v_z\frac{\partial v_\theta}{\partial z}\right)=-\frac{1}{r}\frac{\partial p}{\partial \theta}+\rho g_\theta+\mu\left[\frac{1}{r}\frac{\partial}{\partial r}\left(r\frac{\partial v_\theta}{\partial r}\right)-\frac{v_\theta}{r^2}+\frac{1}{r^2}\frac{\partial^2 v_\theta}{\partial \theta^2}+\frac{2}{r^2}\frac{\partial v_r}{\partial \theta}+\frac{\partial^2 v_\theta}{\partial z^2}\right] \quad (5.30b)$$

$$\rho\left(\frac{\partial v_z}{\partial t}+v_r\frac{\partial v_z}{\partial r}+\frac{v_\theta}{r}\frac{\partial v_z}{\partial \theta}+v_z\frac{\partial v_z}{\partial z}\right)=-\frac{\partial p}{\partial z}+\rho g_z+\mu\left[\frac{1}{r}\frac{\partial}{\partial r}\left(r\frac{\partial v_z}{\partial r}\right)+\frac{1}{r^2}\frac{\partial^2 v_z}{\partial \theta^2}+\frac{\partial^2 v_z}{\partial z^2}\right] \quad (5.30c)$$

5.4 Solutions for Viscous Incompressible Flow

5.4 黏性不可压缩流动的求解

Since the convective acceleration terms are nonlinear, it is difficult to solve the Navier-Stokes equations. In most practical flow cases, accelerated motion of fluid does exist. Thus, the convective acceleration terms are usually important. However, for some cases without convective acceleration, it is possible to obtain the exact solutions. The Navier-Stokes equations can be applied to both laminar and turbulent flows, but for turbulent flow it will be more complicated due to the random velocity fluctuation with respect to time. Thus, the exact solutions can be obtained only for laminar flows.

由于对流加速度项的非线性使纳维-斯托克斯方程的求解变得困难。在大多数实际流动中，确实存在加速运动，因此，对流加速度项通常很重要。然而，对于不存在对流加速度的情况，纳维-斯托克斯方程可以得到精确解。纳维-斯托克斯方程可应用于层流和湍流，但对于湍流流动，速度随时间随机波动，问题变得更为复杂。因此，只有层流流动可以得到精确解。

5.4.1 Steady, Laminar Flow Between Fixed Parallel Plates

5.4.1 固定平板间的定常层流流动

Figure 5.6 shows a flow between two horizontal, infinite parallel plates. In this case fluid flows along the x direction, and the velocity components in the y and z direction are zero, that is, $v_y=0$ and $v_z=0$. Under

两块水平放置的无限大平行平板间的流动如图 5.6 所示。流体沿 x 方向平行于平板流动，y 和 z 方向的速度分量为零，即 $v_y=0$，$v_z=0$。

this condition the continuity equation [Eq. (5.12)] reduce to $\partial v_x/\partial x = 0$. Furthermore, for infinite plates, v_x will not change in the z direction, and for steady flow, $\partial v_x/\partial t = 0$, so that $v_x = v_x(y)$. The gravity shown in Fig. 5.6 is in the negative direction of y axis, that is, $g_x = 0$, $g_y = -g$ and $g_z = 0$. For this flow, the Navier-Stokes equations [Eq. (5.29)] reduce to

这种情况下，式(5.12) 简化为$\partial v_x/\partial x = 0$。另外，对于无限大平板，速度$v_x$在$z$方向上不变，对于定常流动，$\partial v_x/\partial t = 0$，因此，$v_x = v_x(y)$。如图5.6所示的重力方向沿$y$轴负向，即：$g_x = 0$, $g_y = -g$, $g_z = 0$。对于这种流动，纳维-斯托克斯方程[式(5.29)]简化为

$$0 = -\frac{\partial p}{\partial x} + \mu\left(\frac{\partial^2 v_x}{\partial y^2}\right) \quad (5.31)$$

$$0 = -\frac{\partial p}{\partial y} - \rho g \quad (5.32)$$

$$0 = -\frac{\partial p}{\partial z} \quad (5.33)$$

It can be seen that for this particular problem the Navier-Stokes equations reduce to some rather simple equations.

可见，对于这个特殊的问题，纳维-斯托克斯方程变得相当简单。

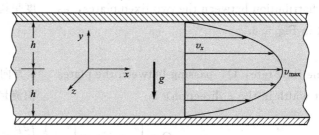

Fig. 5.6　The viscous flow between parallel plates
图5.6　平行平板间的黏性流动

From Eq. (5.33) it can be seen that p is independent of z, so the integration of Eq. (5.32) gives

由式(5.33)可知，p与z无关，对式(5.32)进行积分，得

$$p = -\rho g y + f_1(x) \quad (5.34)$$

which shows that the hydrostatical pressure varies in the y direction. Equation(5.31) can be rewritten as

上式表明，流体静压会沿y方向发生变化。式(5.31)可写为

$$\frac{d^2 v_x}{dy^2} = \frac{1}{\mu}\frac{\partial p}{\partial x}$$

It can be integrated to give

积分后得

$$\frac{dv_x}{dy} = \frac{1}{\mu}\left(\frac{\partial p}{\partial x}\right)y + c_1$$

and integrated again to yield

$$v_x = \frac{1}{2\mu}\left(\frac{\partial p}{\partial x}\right)y^2 + c_1 y + c_2 \quad (5.35)$$

In the integrations the pressure gradient, $\partial p/\partial x$, is treated as constant since it is not a function of y [Eq. (5.34)]. The two integration constants c_1 and c_2 can be determined using the boundary conditions. For example, if the two plates are fixed, $v_x = 0$ when $y = \pm h$ (because of the no-slip condition for viscous fluids). To satisfy this boundary condition, $c_1 = 0$ and $c_2 = -\frac{1}{2\mu}\left(\frac{\partial p}{\partial x}\right)h^2$. Thus, the velocity distribution can be expressed as

$$v_x = \frac{1}{2\mu}\left(\frac{\partial p}{\partial x}\right)(y^2 - h^2) \quad (5.36)$$

Equation (5.36) indicates that the velocity presents a parabolic distribution between the two fixed plates, as illustrated in Fig. 5.6.

The volume flow rate, Q, passing between the plates (for a unit width in the z direction) is

$$Q = \int_{-h}^{h} v_x \, dy = \int_{-h}^{h} \frac{1}{2\mu}\left(\frac{\partial p}{\partial x}\right)(y^2 - h^2) \, dy$$

$$Q = -\frac{2h^3}{3\mu}\left(\frac{\partial p}{\partial x}\right) \quad (5.37)$$

Since the pressure decreases in the direction of flow, the pressure gradient $\partial p/\partial x$ is negative. If Δp represents the pressure drop between two points with a distance l, then

$$\frac{\Delta p}{l} = -\frac{\partial p}{\partial x}$$

and Eq. (5.37) can be expressed as

$$Q = \frac{2h^3 \Delta p}{3\mu l} \quad (5.38)$$

再次积分后得

因为压力梯度$\partial p/\partial x$ 不是 y 的函数，积分时可以将其看作常数。积分常数c_1和c_2由边界条件进行确定。例如：如果两平板固定，当$y = \pm h$时，$v_x = 0$（黏性流体无滑移边界条件）。为了满足该边界条件，$c_1 = 0$，$c_2 = -\frac{1}{2\mu}\left(\frac{\partial p}{\partial x}\right)h^2$。因此，速度分布可表达为

式(5.36) 表明两固定平板间的速度呈抛物线分布，如图 5.6 所示。

z 方向单位宽度上通过平板间的体积流量为

由于压力沿流动方向降低，压力梯度$\partial p/\partial x$为负值。如果Δp 表示相距为 l 的两点之间的压力降，那么

式(5.37) 可写为

Because the mean velocity $v = Q/(2h)$, Eq. (5.38) becomes

$$v = \frac{h^2 \Delta p}{3\mu l} \quad (5.39)$$

Since the maximum velocity, v_{max}, occurs at $y=0$, from Eq. (5.36)

$$v_{max} = -\frac{h^2}{2\mu}\left(\frac{\partial p}{\partial x}\right) = \frac{h^2 \Delta p}{2\mu l}$$

so that

$$v_{max} = \frac{3}{2} v \quad (5.40)$$

5.4.2 Steady, Laminar Flow in Circular Tubes

Figure 5.7 illustrates the flow through a horizontal circular tube with radius R. For convenience cylindrical coordinates are applied for this case. It is assumed to be a one-dimensional flow along the z direction, so that $v_r = 0$ and $v_\theta = 0$, and from the continuity equation [Eq. (5.15)] $\partial v_z/\partial z = 0$. Also, for this steady and axisymmetric flow, v_z is independent of t or θ, so the velocity, v_z, is only a function of the radial coordinate, that is, $v_z = v_z(r)$. The components of acceleration of gravity in r and θ directions (g_r and g_θ) are $-g\sin\theta$ and $-g\cos\theta$, respectively, where θ is the angle between the radial direction and the horizontal plane. Under these conditions, the Navier-Stokes equations [Eq. (5.30)] are simplified as

$$0 = -\rho g \sin\theta - \frac{\partial p}{\partial r} \quad (5.41)$$

$$0 = -\rho g \cos\theta - \frac{1}{r}\frac{\partial p}{\partial \theta} \quad (5.42)$$

$$0 = -\frac{\partial p}{\partial z} + \mu\left[\frac{1}{r}\frac{\partial}{\partial r}\left(r\frac{\partial v_z}{\partial r}\right)\right] \quad (5.43)$$

Integration of Eq. (5.41) and Eq. (5.42) gives

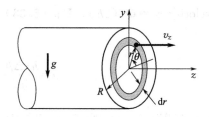

Fig. 5.7 The viscous flow in a circular tube
图 5.7 圆管内的黏性流动

$$p = -\rho g(r\sin\theta) + f_1(z)$$

which is 即

$$p = -\rho g y + f_1(z) \quad (5.44)$$

Equation (5.44) indicates that the z component of the pressure gradient, $\partial p/\partial z$, is not a function of r or θ.

式(5.44) 表明，z 方向上的压力梯度 $\partial p/\partial z$ 不是 r 或 θ 的函数。

The equation of motion in the z direction [Eq. (5.43)] can be written as

z 方向上的运动方程[式(5.43)]可以写成

$$\frac{1}{r}\frac{\partial}{\partial r}\left(r\frac{\partial v_z}{\partial r}\right) = \frac{1}{\mu}\frac{\partial p}{\partial z}$$

Since $\partial p/\partial z$ is constant, it can be integrated to give

由于 $\partial p/\partial z$ 为常数，积分后得

$$r\frac{\partial v_z}{\partial r} = \frac{1}{2\mu}\left(\frac{\partial p}{\partial z}\right)r^2 + c_1$$

and integrated again to yield 再次积分后得

$$v_z = \frac{1}{4\mu}\left(\frac{\partial p}{\partial z}\right)r^2 + c_1\ln r + c_2 \quad (5.45)$$

Since v_z is finite at the center of the tube ($r=0$), it follows that $c_1=0$ [since $\ln(0)=-\infty$]. At the wall boundary ($r=R$) $v_z=0$, so that

由于在圆管的中心 ($r=0$) 速度为有限值，故 $c_1=0$ [因为 $\ln(0)=-\infty$]。在壁面处 ($r=R$)，$v_z=0$，所以

$$c_2 = -\frac{1}{4\mu}\left(\frac{\partial p}{\partial z}\right)R^2$$

and the velocity distribution can be expressed as 速度分布可表达为

$$v_z = \frac{1}{4\mu}\left(\frac{\partial p}{\partial z}\right)(r^2 - R^2) \quad (5.46)$$

It can be seen that the velocity distribution is parabolic at any cross section.

可见，在任何横截面上，速度呈抛物线分布。

To calculate the volume flow rate, Q, passing through the tube, the differential annulus shown in Fig. 5.7 is extracted, and v_z is constant on this annulus. Then the volume flow rate through the differential area $dA = (2\pi r)dr$ is

为了计算通过圆管的体积流量 Q，取如图 5.7 所示的微元环，微元环上速度 v_z 为常数，通过面积 $dA = (2\pi r)dr$ 的体积流量为

$$dQ = v_z(2\pi r)dr$$

and therefore

因此

$$Q = 2\pi \int_0^R v_z r\, dr \quad (5.47)$$

Substituting Eq. (5.46) into Eq. (5.47) and integrating give

将式(5.46)代入式(5.47)，积分后得

$$Q = -\frac{\pi R^4}{8\mu}\left(\frac{\partial p}{\partial z}\right) \quad (5.48)$$

If Δp represents the pressure drop over a length, l, along the tube, then

如果 Δp 表示沿圆管相距长度为 l 的两点之间的压力降，那么

$$\frac{\Delta p}{l} = -\frac{\partial p}{\partial z}$$

and Eq. (5.48) can be expressed as

式(5.48)可表示为

$$Q = \frac{\pi R^4 \Delta p}{8\mu l} \quad (5.49)$$

Since the mean velocity $v = Q/(\pi R^2)$, Eq. (5.49) becomes

因为平均速度 $v = Q/(\pi R^2)$，式(5.49)变为

$$v = \frac{R^2 \Delta p}{8\mu l} \quad (5.50)$$

The maximum velocity v_{max} occurs at the center of the tube, from Eq. (5.46)

最大速度 v_{max} 出现在圆管的中心，由式(5.46)得

$$v_{max} = -\frac{R^2}{4\mu}\left(\frac{\partial p}{\partial z}\right) = \frac{R^2 \Delta p}{4\mu l} \quad (5.51)$$

so that

因此

$$v_{max} = 2v$$

Exercises

习题

5.1 The three components of velocity in a flow field

5.1 流场中三个方向的速度分量

are given by $v_x = x^2$, $v_y = -2xy$, $v_z = x + y$. (a) Determine the volumetric dilatation rate and interpret the results. (b) Is this an irrotational flow field?

5.2 What combination of constants, a, b, c and d can the velocity components ($v_x = ax + by$, $v_y = cx + dy$, $v_z = 0$) be used to describe an incompressible flow field? For incompressible fluids, $\nabla \cdot \mathbf{v} = 0$.

5.3 The velocity components of an incompressible, two-dimensional flow field are given by $v_x = 2xy$, $v_y = x^2 - y^2$. Prove that the flow is irrotational and satisfies conservation of mass.

5.4 A layer of viscous liquid of constant thickness flows steadily down an infinite, inclined plane as shown in Fig. 5.8. Assume that the flow is laminar and air resistance is negligible. Determine the velocity distribution in the layer by means of the Navier-Stokes equations.

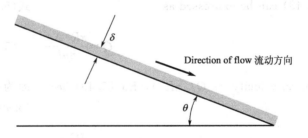

Fig. 5.8 Figure for the exercise 5.4
图 5.8 习题 5.4 图

5.5 A viscous, incompressible fluid flows steadily between the two infinite, vertical, parallel plates as shown in Fig. 5.9. Assume that the flow is laminar. Make use of the Navier-Stokes equation to determine: (a) the velocity distribution between the two plates, (b) the flow rate per unit width.

5.6 An incompressible, viscous fluid is placed between horizontal, infinite, parallel plates as shown in Fig. 5.10. The two plates move in opposite directions

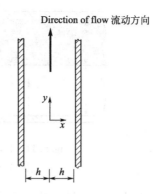

Fig. 5.9　Figure for the exercise 5.5
图 5.9　习题 5.5 图

with constant velocities, v_1 and v_2. Assume the flow is laminar. The pressure gradient in the x direction is zero, and the only body force is due to the fluid weight. Derive an expression for the velocity distribution between the plates by use of the Navier-Stokes equations.

假设流动为层流流动，x 方向的压力梯度为零，并考虑流体重力。试采用纳维-斯托克斯方程推导两板间的速度分布表达式。

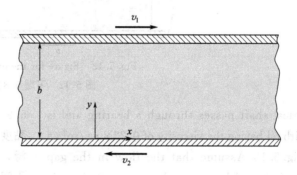

Fig. 5.10　Figure for the exercise 5.6
图 5.10　习题 5.6 图

5.7　Two immiscible, incompressible, viscous fluids are contained between two infinite, horizontal, parallel plates as shown in Fig. 5.11. The bottom plate is fixed and the upper plate moves with a constant velocity v. Assume the flow is laminar, and there is no pressure gradient in the x direction. Determine velocity distribution between the two plates and the velocity at the interface.

5.7　两块无限大水平放置的平行平板间充满两种不相容的不可压缩黏性流体如图 5.11 所示。下板固定，上板以速度 v 运动。假设流动为层流流动，x 方向上没有压力梯度。试确定两板间的速度分布及交界面上的速度大小。

5.8　The viscous, incompressible flow between the parallel plates shown in Fig. 5.12 is caused by both the motion of the bottom plate and a pressure gradient,

5.8　如图 5.12 所示的平行平板间的黏性不可压缩流动由下板的运动和压力梯度 $\partial p/\partial x$ 引起。假设

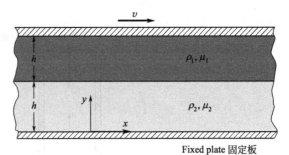

Fig. 5.11　Figure for the exercise 5.7

图 5.11　习题 5.7 图

$\partial p/\partial x$. Assume the flow is laminar. Use the Navier-Stokes equations to derive an expression for the velocity distribution between the plates.

流动为层流流动。试采用纳维-斯托克斯方程推导两板间的速度分布。

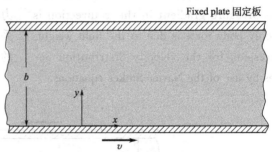

Fig. 5.12　Figure for the exercise 5.8

图 5.12　习题 5.8 图

5.9　A vertical shaft passes through a bearing and is lubricated with oil having the viscosity of 0.22N·s/m² as shown in Fig. 5.13. Assume that the flow in the gap between the shaft and bearing is laminar and there is no pressure gradient in the direction of flow. Estimate the torque required to overcome viscous resistance when the shaft is turning at 100 rpm.

5.9　一根垂直的轴穿过轴承，靠黏度为 0.22N·s/m² 的油进行润滑，如图 5.13 所示。假设轴和轴承间隙内的流动为层流，在流动方向上没有压力梯度。当轴的转速为 100rpm 时，试估算克服黏性阻力需要多大的扭矩。

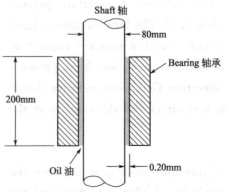

Fig. 5.13　Figure for the exercise 5.9

图 5.13　习题 5.9 图

Chapter 6 Similitude and Dimensional Analysis
第6章 相似理论和量纲分析

The theoretical methods introduced in the preceding chapters can be used to solve many engineering problems, but there still exist other problems that cannot be solved analytically and may depend on experimental investigations. A perfect solution to any problem is usually obtained by analysis and experimental methods jointly. Therefore engineers should be familiar with the experimental approaches to fluid mechanics problems so that they can execute an experiment or make use of the experimental data.

The rig established in laboratory is usually thought of as model while the instrument which is developed based on the experimental result is prototype. Obviously model and prototype should have certain similar characteristics so that the experimental data can be used to guide the design of prototype. The concept of similitude is often used to judge measurements made on a laboratory system can be used to describe the behaviors of prototype system.

How to ensure the flow in model system is similar to that in prototype system? It is necessary to establish physical relationships between model and prototype.

6.1 Similitude

For two geometrically similar systems, if the ratio of a physical quantity in one system to the corresponding quantity in another system is certain, it can be thought that the flows in the two systems are similar. Then we can interpret the flow characteristics of one flow through

前文介绍的理论方法可用于解决很多流体力学问题,但仍有部分问题的解决不能采用解析方法,只能依赖于实验研究。解决某一问题最完善的做法是同时采用理论分析和实验研究。因此,工程人员有必要掌握研究流体力学问题的实验方法,并能利用实验数据。

通常把在实验室建立的装置称为模型,而根据实验结果建立的装置称为原型。显然,为了使实验结果能指导原型设计,必须保证实验模型与原型具有一定的相似特征。相似理论的概念可用于判断实验系统(模型)能否用于描述原型的特性。

如何保证模型与原型的流动相似呢?这需要建立模型与原型的物理关系。

6.1 相似理论

对于两个几何相似的系统,如果对应点上的物理量成比例,可以认为两个系统中的流动是相似的。这样就可以通过其中一个系统来研究另一个系统。流动相似包括

another one. Flow similitude compromises geometrical similarity, kinematic similarity and dynamic similarity.

(1) Geometrical similarity

Hydraulic models may be either true or scale models. Model reproduces features of the prototype but at a scale—that is they are geometrically similar. In two geometrically similar models, the length ratios between the points are equal everywhere.

Actually it is very difficult to realize the perfect geometrical similarity. For example, it is impossible to scale up or down the rough surface from prototype to model.

(2) Kinematic similarity

Kinematic similarity is the similarity of time as well as geometry. It requires model and prototype to meet that: ①the paths of moving particles are geometrically similar, ② the ratios of the velocities of corresponding particles equal to a constant. This has the consequence that streamline patterns are the same.

(3) Dynamic similarity

Dynamic similarity first should be geometrical and kinematic similarity, then the forces on corresponding points of model and prototype are in the same direction and the ratio of their values is certain. In fluid mechanics, dynamic similarity usually includes both force and heat similarities. The force and heat quantities of fluid mechaics include gravity, pressure difference force, viscous stress, surface tension, inertia force, buoyancy, solid surface heat release, conductive heat, convective heat and radiation and viscous dissipation, etc.

Geometrical similarity, kinematic similarity and dynamic similarity are dependent upon each other and involved in physical similarity. The physical similarity

几何相似、运动相似和动力相似。

(1) 几何相似

水力模型可以是全尺寸模型或比例模型。模型应该按一定比例再现原型的特征，这就是几何相似。两个几何相似的模型中，对应点之间的长度比处处相等。

完全的几何相似很难实现。例如，等比例放大或缩小原型的粗糙表面几乎无法实现。

(2) 运动相似

运动相似是指在几何相似的基础上，时间相似。运动相似的模型和原型要求：① 两个系统对应点的运动轨迹要几何相似；② 对应点的速度比值是常数。这使得模型和原型的流线形式一致。

(3) 动力相似

动力相似的系统首先应是几何相似和运动相似的，然后要求对应点上力的方向相同，且比值一定。流体力学中，动力相似包括力和热的相似。一般流体力学问题中包括：重力、压差力、黏性应力、表面张力、惯性力、浮力、固壁放热、流体之间传导热、对流热与辐射热、流体黏性耗散热等。

几何相似、运动相似和动力相似彼此联系又相互制约，它们统一在物理相似中。根据物理相似定义，

requires all the dimensionless characteristic parameters of model should equal to that of prototype one by one.

6.2 Similarity Laws

We have at least two methods to obtain similarity laws. One is the dimensional analysis which is a straightforward approach to modeling. The other is to develop the similarity laws from the governing equations (N-S equations). The former is a relatively simple method that only requires to know the variables that influence the physical process, which will be mentioned in section 6.3. To illustrate the procedure of the later method, we consider an incompressible viscous flow governed by the N-S equations.

For simplicity, the gravity is the only body force exerting on fluid particles. The subscripts p and m represent prototype and model respectively. For prototype, the N-S equation for the z component is in the form

$$\frac{\partial v_{pz}}{\partial t_p}+v_{px}\frac{\partial v_{pz}}{\partial x_p}+v_{py}\frac{\partial v_{pz}}{\partial y_p}+v_{pz}\frac{\partial v_{pz}}{\partial z_p}=-g_{pz}-\frac{1}{\rho_p}\frac{\partial p_p}{\partial z_p}+\frac{\mu_p}{\rho_p}\left(\frac{\partial^2 v_{pz}}{\partial x_p^2}+\frac{\partial^2 v_{pz}}{\partial y_p^2}+\frac{\partial^2 v_{pz}}{\partial z_p^2}\right) \quad (6.1)$$

and for model, it holds

$$\frac{\partial v_{mz}}{\partial t_m}+v_{mx}\frac{\partial v_{mz}}{\partial x_m}+v_{py}\frac{\partial v_{mz}}{\partial y_m}+v_{pz}\frac{\partial v_{mz}}{\partial z_m}=-g_{mz}-\frac{1}{\rho_m}\frac{\partial p_m}{\partial z_m}+\frac{\mu_m}{\rho_m}\left(\frac{\partial^2 v_{mz}}{\partial x_m^2}+\frac{\partial^2 v_{mz}}{\partial y_m^2}+\frac{\partial^2 v_{mz}}{\partial z_m^2}\right) \quad (6.2)$$

Considering the similitude between prototype and model, the corresponding parameters in Eq. (6.1) and Eq. (6.2) must comply with the following relationships.

① The geometrical similarity gives

$$C_l = \frac{x_p}{x_m} = \frac{y_p}{y_m} = \frac{z_p}{z_m} \quad (6.3)$$

② The kinematic similarity gives

$$C_v = \frac{v_{px}}{v_{mx}} = \frac{v_{py}}{v_{my}} = \frac{v_{pz}}{v_{mz}} \quad \text{and} \quad C_t = \frac{t_p}{t_m} \quad (6.4)$$

③ The dynamics similarity gives

6.2 相似准则

有两种方法可以得到相似准则，一是采用量纲分析法（直接建模方法），二是从流动控制方程（N-S方程）中分析得到。前者相对简单，仅需要知道影响物理过程的变量，这一方法将在6.3节介绍。本节通过考虑不可压缩黏性流动的N-S方程来说明相似准则的推导方法。

简便起见，质量力仅考虑重力。下标p和m分别表示原型和模型。原型沿z坐标方向的N-S方程表示为

对于模型有

由于模型和原型是相似的，式(6.1)、式(6.2)中的各个参数应具有如下关系。

① 根据几何相似，有

② 根据运动相似，有

③ 根据动力相似，有

$$C_p = \frac{p_p}{p_m} \quad (6.5)$$

Furthermore, consider the density and viscosity similarities

此外，根据密度和黏度相似，有

$$C_\rho = \frac{\rho_p}{\rho_m} \text{ and } C_\mu = \frac{\mu_p}{\mu_m} \quad (6.6)$$

Substitution of Eq. (6.3), Eq. (6.4), Eq. (6.5) and Eq. (6.6) into Eq. (6.1) gives another form of differential equation for prototype

将式(6.3)、式(6.4)、式(6.5)和式(6.6)代入式(6.1)，得

$$\frac{C_v}{C_t}\frac{\partial v_{mz}}{\partial t_m} + \frac{C_v^2}{C_l}\left(v_{mx}\frac{\partial v_{mz}}{\partial x_m} + v_{py}\frac{\partial v_{mz}}{\partial y_m} + v_{pz}\frac{\partial v_{mz}}{\partial z_m}\right) =$$

$$-C_g g_{zm} - \frac{C_p}{C_l C_\rho}\frac{1}{\rho_m}\frac{\partial p_m}{\partial z_m} + \frac{C_v C_\mu}{C_l^2 C_\rho}\frac{\mu_m}{\rho_m}\left(\frac{\partial^2 v_{mz}}{\partial x_m^2} + \frac{\partial^2 v_{mz}}{\partial y_m^2} + \frac{\partial^2 v_{mz}}{\partial z_m^2}\right) \quad (6.7)$$

Comparing Eq. (6.2) and Eq. (6.7), if

比较式(6.2)和式(6.7)，如果

$$\frac{C_v}{C_t} = \frac{C_v^2}{C_l} = C_g = \frac{C_p}{C_l C_\rho} = \frac{C_v C_\mu}{C_l^2 C_\rho} \quad (6.8)$$

the coefficients in Eq. (6.7) can be omitted and then it is in the same form to the equation of model. This means, for prototype and model, if the relationships shown by Eq. (6.8) is valid and the boundary conditions are also similar, the flow in prototype is similar to that in model because they are governed by the same motion equations.

则式(6.7)中的各项系数可被约去，模型的方程与原型一致。亦即，对于模型和原型，如果式(6.8)成立且边界条件相似，则可认为模型和原型的流动相似，因为两者拥有同一套运动方程。

Divided by C_v^2/C_l for each term, Eq. (6.8) becomes

式(6.8)中各项除以C_v^2/C_l，有

$$\frac{C_v C_l C_\rho}{C_\mu} = \frac{C_p}{C_\rho C_v^2} = \frac{C_v^2}{C_l C_g} = \frac{C_l}{C_t C_v} = 1 \quad (6.9)$$

Considering the first term in Eq. (6.9), $C_v C_l C_\rho/C_\mu = 1$, the physical equation of the characteristic length, l, velocity, v, viscosity, μ, and density, ρ, for prototype and model should be in the form

由第1项，即$C_v C_l C_\rho/C_\mu = 1$，可以得到模型和原型中特征长度l，速度v，黏度μ和密度ρ之间的关系，即

$$\frac{l_p v_p \rho_p}{\mu_p} = \frac{l_m v_m \rho_m}{\mu_m} \quad (6.10)$$

Equation (6.10) shows that the value of the dimensionless parameter group, $vl\rho/\mu$ (the Reynolds number, Re), for prototype should equal to that of the model. Similarly, other three dimensionless groups, such as the Euler number ($Eu = p/\rho v^2$), the Froude number ($Fr = v^2/gl$) and the Strouhal number ($St = l/vt$), can also be obtained. Their physical explanations are listed in Table 6.1.

式(6.10)说明模型中的无量纲参数 $vl\rho/\mu$（即雷诺数 Re）应与原型相等。类似的，其他3个无量纲参数，如欧拉数（$Eu = p/\rho v^2$），弗劳德数（$Fr = v^2/gl$），斯特劳哈数（$St = l/vt$）也应分别相等，它们的物理意义参见表6.1。

Table 6.1 Similarity laws obtained by N-S equations analysis
表6.1 由N-S方程分析得到的相似准则

Dimensionless groups 无量纲组合	Definition 定义	Physical explanation 物理意义	Types of application 适用场合
$Re = vl\rho/\mu$	Reynolds number（雷诺数），Re	viscous force vs. inertia force 黏性力比惯性力	pipe flow, drag force problem, boundary layer problem 管内流动、曳力问题、边界层流动
$Eu = p/\rho v^2$	Euler number（欧拉数），Eu	pressure vs. inertia force 压力比惯性力	water hammer, vacuole effect and cavity drag in pipe flow 水击现象、空泡效应、管内流动的空泡阻力
$Fr = v^2/gl$	Froude number（弗劳德数），Fr	inertia force vs. gravity 惯性力比重力	tide, weir flow, fluctuation, wave drag 潮汐、堰流、波动、波阻
$St = l/vt$	Strouhal number（斯特劳哈数），St	force induced by velocity change vs. inertia force 由速度变化引起的力比惯性力	blade flow, propeller flow 桨叶流动、螺旋桨流动

The **Reynolds Number** can be used as a criterion to distinguish between laminar and turbulent flow. It is a measure of the ratio of the inertia force to the viscous force exerting on an element of fluid.

雷诺数可用来判定流动状态，即层流或是湍流。雷诺数的物理意义在于其表明了作用于流体单元上的惯性力和黏性力的比值。

The **Euler Number** can be interpreted as a measure of the ratio of pressure force to inertial force, where the pressure is characteristic pressure in the flow field.

欧拉数可以理解为压力和惯性力的比值，其中压力指的是流场的特征压力。

The **Froude Number** contains the acceleration of gravity, g, which means the acceleration of gravity becomes an important factor in the problem of interest. The Froude number is a measure of the ratio of the inertia force to the weight of the element. It should be considered in problems involving flows with free surfaces since gravity principally affects the flow.

弗劳德数中包含了重力加速度 g，说明涉及弗劳德数的问题中重力加速度影响显著，它实际上是作用于流体单元上的惯性力和重力的比值。在有自由表面的流动问题中需要考虑弗劳德数。

The **Strouhal Number** is an important dimensionless parameter in unsteady, oscillating flow problems. It is a measure of the ratio of inertial forces due to the unsteadiness of the flow (local acceleration) to the inertial forces due to changes in velocity while fluid element transports (convective acceleration) in the flow field.

6.3 Dimensional Analysis

6.3.1 Dimensional Homogeneity Principle

As introduced in Chapter 1, we can represent the dimensions of any physical quantity of interest with L, M, T and Θ. For example, the dimension of velocity is $[l][t]^{-1}$ or LT^{-1} and the dimension of force is $[m][l][t]^{-2}$ or MLT^{-2}. In this chapter only L, M and T are concerned.

If the dimension of a physical quantity, such as velocity or force, can be expressed by the group of other dimensions, it is a dependent dimension.

The physical quantities with no unit attached are termed nondimensional quantities, and their magnitudes are not related to unit system. It should be noted that, in physics, the plane angle is defined as the ratio of arc length to radius of curvature, and the solid angle is defined as the result of surface area divided by the square of radius of curvature. Therefore both plane angle and solid angle are dimensionless quantities.

In a dimensional analysis, it is of great importance to distinguish between the variables with independent dimensions and that with dependent dimensions. In general fluid mechanics problems, the number of variables with independent dimensions is no larger than four. In fluid kinematics problems, only two variables with independent dimensions are involved; in incompressible fluid dynamics problem without heat transfer, the number is three; in other general fluid dynamics problems (not including electromagnetic fluid mechanics), four variables with independent dimensions are involved.

在非稳定的周期性流动问题中需要考虑**斯特劳哈数**。斯特劳哈数是由流动的不稳定性（局部加速度）引起的惯性力与质点迁移（对流加速度）时速度变化引起的惯性力的比值。

6.3 量纲分析

6.3.1 量纲和谐原理

如第1章所介绍的，所有物理量的量纲均可用L、M、T和Θ的组合来表示。例如，速度的量纲可表示为$[l][t]^{-1}$或LT^{-1}，力的量纲是$[m][l][t]^{-2}$或MLT^{-2}。本章只涉及L、M和T。

若一个物理量的量纲能够用其他物理量的量纲组合来表示，则该物理量的量纲不独立。

没有单位的物理量称为无量纲量，其大小与选择的单位系统无关。角度在物理学中是以弧度表示的，平面角定义为对应的弧长除以曲率半径，立体角定义为对应的曲面面积除以曲率半径的平方，都没有单位，所以角度是无量纲量。

进行量纲分析时，区分出独立量纲量和非独立量纲量是重要的第一步。在一般流体力学问题中，独立量纲量的数目不大于4。对于流体运动学问题，独立量纲量只有2个；对于不可压缩流体动力学问题，不考虑传热时，独立量纲量为3个；其他一般的流体动力学问题（不包括电磁流体力学）独立量纲量为4个。

If it is possible to identify the factors involved in a physical situation, dimensional analysis can help form a relationship among them. The resulting expressions from a dimensional analysis may not rigorous at first sight appear but these qualitative results converted to quantitative forms can be used to obtain any unknown factors from experimental analysis.

只要能确定一个物理过程的影响因素，就可通过量纲分析建立这些因素之间的联系。量纲分析得到的表达式并不严密，但通过实验分析可把定性结果转化为定量结果，可以发现隐藏的影响因素。

Any equation describing a physical situation will be true only if both sides have the same dimensions. That is it must be dimensionally homogenous. For example, the flow rate equation for a flow over a rectangular weir (Fig. 6.1) is

用于描述物理现象的方程正确的首要条件是两边的量纲一致，这称为量纲和谐。例如，溢流堰（图 6.1）的流量方程为

$$Q = \frac{2}{3} b \sqrt{2g} h^{\frac{3}{2}} \qquad (6.11)$$

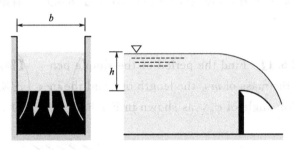

Fig. 6.1　A overflow weir model
图 6.1　溢流堰模型

The SI unit of the left hand side are m^3/s, so the unit of the right hand side must be the same. Writing the equation with only the SI units gives

式(6.11)左边的单位是 m^3/s，式(6.11)右边的单位也应是 m^3/s。式(6.11)左右两边的单位可以写成

$$m^3 \cdot s^{-1} = m \cdot m^{0.5} \cdot s^{-1} \cdot m^{1.5}$$

Obviously the units of the two sides of the equation are consistent. To be stricter, it is the dimensions that must be consistent.

显然，式(6.11)两边的单位是一致的。严格地说，方程两边一致的是量纲。

6.3.2　The Rayleigh Method

If the factors involved in a problem present the relationship of the production of power exponents of all variables, the Rayleigh method (an indicial method) can be used

6.3.2　瑞利法

瑞利分析法（指数法）的前提条件是影响流动现象的变量之间的函数关系是幂函数乘积的形式，求

to obtain the powers by the following 5 steps.

Step 1   Make sure this is a problem that can be applied by the Rayleigh method;

Step 2   Select the physical quantities in the problem;

Step 3   Establish the exponential equation according to the dimensional homogeneity principle. Usually a dimensionless constant to be determined should be involved in the equation;

Step 4   Establish the equations about all the exponents and solve each exponent;

Step 5   Determine the dimensionless constant using an experimental method.

The result of performing the Rayleigh dimensional analysis on a physical problem is a single equation which relates all of the physical factors involved to one another.

【EXAMPLE 6.1】 Find the period of the simple pendulum with the mass of m, the length of l and the initial oscillation angle of φ_0, as shown in Fig. 6.2.

解过程包括以下 5 步。

第 1 步   确定这是一个瑞利法可以解决的问题；

第 2 步   找出相关的物理量；

第 3 步   写出各物理量幂指数乘积形式的方程，通常应包含一个待定的无量纲常数；

第 4 步   列出关于各幂指数的方程，求解各幂指数；

第 5 步   用试验方法确定待定常数。

采用瑞利量纲分析法得到的是一个简单的方程，该方程建立了所有影响因素之间的关系。

【例题 6.1】 如图 6.2 所示，单摆的质量为 m，长度为 l，初始摆角为 φ_0，试求单摆的周期。

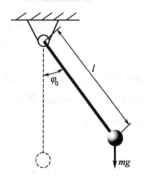

Fig. 6.2   A single pendulum model
图 6.2   单摆模型

SOLUTION   The vibration of the pendulum is initiated by the gravity, hence the acceleration of gravity, g, is an important quantity in this problem. Besides, the period T_p may also be affected by m, l and φ_0, so T_p can be expressed as

解   单摆的振动主要是由重力引起的，因此重力加速度是一个重要因素。此外，单摆的周期 T_p 可能还与 m、l 和 φ_0 有关。于是

$$T_p = f(m, l, \varphi_0, g)$$

Notice that φ_0 is a dimensionless quantity and then

in the dimension form

注意到 φ_0 是无量纲量，因此

$$[T_p]=[m]^a[l]^b[g]^c$$

量纲形式可写为

$$T=(M)^a(L)^b(LT^{-2})^c$$

Comparison of the dimensions of the two hand sides of the equation gives

比较等式两边的量纲，可得

For dimension M: $0=a$

For dimension L: $0=b+c$

For dimension T: $1=-2c$

Solutions of the exponents are

求得各指数为

$$a=0, b=1/2, c=-1/2 \text{ and } T_p=k\sqrt{l/g}$$

where k is a dimensionless constant to be determined according to experiment result, and $k=2\pi$.

式中，k 为无量纲常数，可根据实验确定，$k=2\pi$。

6.3.3 The Buckingham's Π Theorems

6.3.3 白金汉姆 Π 定理

Assuming Π is a dimensionless variable, a quantity with dependent dimension, a, and some quantities with independent dimension $a_1, a_2, \cdots, a_k$ hold the relationship as

假定 Π 为无量纲量，则量纲不独立的物理量 a 和量纲独立的物理量 $a_1, a_2, \cdots, a_k$ 之间有如下关系

$$\Pi = \frac{a}{a_1^{m_1} a_2^{m_2} \cdots a_k^{m_k}} \quad (6.12)$$

In any physical situation, a quantity to be determined, b, is related to some known variables with dimensions, $a_1, a_2, \cdots, a_n$ (n is the number of the dimensional variables), and other known variables without dimensions, $c_1, c_2, \cdots, c_s$ (s is the number of the dimensionless variables), where $a_1, a_2, \cdots, a_k$ ($k \leqslant n$) are variables with independent dimension. Thereby, $b=f_b(a_1, a_2, \cdots, a_n, c_1, c_2, \cdots, c_s)$. These variables can be grouped into some dimensionless terms and then

任何一个物理现象中，如有一个待定量 b，已知它与 n 个有量纲的已知量 $a_1, a_2, \cdots, a_n$ 以及 s 个无量纲的已知量 $c_1, c_2, \cdots, c_s$ 有关，并已知 $a_1, a_2, \cdots, a_k$ ($k \leqslant n$) 为独立量纲量，则有 $b=f_b(a_1, a_2, \cdots, a_n, c_1, c_2, \cdots, c_s)$。这些变量可以组成一些无量纲项，于是

$$\Pi = f(\Pi_1, \Pi_2, \cdots, \Pi_{n-k}, c_1, c_2, \cdots, c_s) \quad (6.13)$$

where Π presents a dimensionless group and Π_j is the dimensionless group of some known variables, and they hold

$$\Pi = \frac{b}{a_1^{m_1} a_2^{m_2} \cdots a_k^{m_k}}, \quad \Pi_j = \frac{a_{k+j}}{a_1^{m_{1j}} a_2^{m_{2j}} \cdots a_k^{m_{kj}}} (j=1,2,\cdots,n-k)$$

where m_1, m_2, $\cdots$, m_k; m_{1j}, m_{2j}, $\cdots$, m_{kj} are rational numbers. Through the Π theorems, the dimensional function of $b = f_b(a_1, a_2, \cdots, a_n, c_1, c_2, \cdots, c_s)$ is simplified to be the dimensionless function of $\Pi = f(\Pi_1, \Pi_2, \Pi_{n-k}, c_1, c_2, \cdots, c_s)$, so the number of variables is reduced from $(n+s)$ to $(n-k+s)$. Thereby the difficulty in the problem is lowered obviously, for both theoretical and experimental approaches.

【EXAMPLE 6.2】 Determine the velocity of a steady incompressible pipe flow, by taking the density, ρ, viscosity, μ, and characteristic length, l, as known conditions and ignoring the gravity.

SOLUTION The flow velocity is related to ρ, μ, l and the pressure difference Δp between the two ends of the pipe. Thereby

$$[v] = [l]^{a_1} [\Delta p]^{b_1} [\rho]^{c_1}$$

Writing in the form of dimension gives

$$LT^{-1} = (L)^{a_1} (ML^{-1}T^{-2})^{b_1} (ML^{-3})^{c_1}$$

Dimensional analysis of the equation gives

$$a_1 = 0, \quad b_1 = 1/2, \quad c_1 = -1/2$$

Then a dimensionless variable Π_v is be obtained

$$\Pi_v = \frac{v}{\sqrt{\Delta p / \rho}}$$

and it can be rewritten as

$$\Pi_v = \frac{\Delta p}{\rho v^2} = Eu \text{ (the Euler number)}$$

Similarly, if we want to determine the viscosity of a fluid using the pipe flow experiment, the physical relationship can be expressed as

同理，若要通过管内流动试验确定某种流体的黏度，则其中的物理关系可表示为

$$[\mu]=[l]^{a_2}[\Delta p]^{b_2}[\rho]^{c_2}$$

Then 有

$$ML^{-1}T^{-1}=(L)^{a_2}(ML^{-1}T^{-2})^{b_2}(ML^{-3})^{c_2}$$

The solutions are $a_2=1$, $b_2=1/2$, $c_2=1/2$, and another dimensionless quantity Π_μ can be obtained

解得 $a_2=1$，$b_2=1/2$，$c_2=1/2$，可得另一无量纲组合 Π_μ 为

$$\Pi_\mu = \frac{\mu}{l\sqrt{\rho \Delta p}}$$

Considering $Eu=\frac{\Delta p}{\rho v^2}$, Π_μ can be rewritten as $\Pi_\mu = \frac{\mu}{l\sqrt{\rho \rho v^2}} = \frac{\mu}{l\rho v}$ or $\Pi_\mu = \frac{l\rho v}{\mu} = Re$ (the Reynolds number). Applying the Π theorems gives

因为 $Eu=\frac{\Delta p}{\rho v^2}$，$\Pi_\mu$ 可以表示为 $\Pi_\mu = \frac{\mu}{l\sqrt{\rho \rho v^2}} = \frac{\mu}{l\rho v}$ 或 $\Pi_\mu = \frac{l\rho v}{\mu} = Re$（即雷诺数）。根据 Π 定理，有

$$Eu=f(Re) \quad \text{i. e.} \quad \frac{\Delta p}{\rho v^2}=f\left(\frac{l\rho v}{\mu}\right)$$

In this case, the related five variables are organized as two nondimensional groups of variables. As a result, the number of variables is reduced and we now only need to treat two. For example, if we want to double the value Re, we can either double the characteristic length (l) or double the velocity (v), and the two ways will lead to the same result.

本例中，影响过程的 5 个变量被归结为 2 个无量纲的组合。因而试验过程要考虑的 5 个因素减少为 2 个。例如要使 Re 加倍，可将特征尺寸变为 2 倍或速度变为 2 倍，对试验结果的影响是一样的。

Dimensional analysis can be used to obtain similarity laws, and it is a simple and straightforward approach. It should be recognized that omission of any important variable may cause serious errors in model design.

采用量纲分析可获得相似准则，该方法简单、直接。但若遗漏了某些重要变量，将导致模型设计发生严重错误。

6.3.4 Application of the Buckingham's Π Theorems

6.3.4 Π 定理的应用

Compared with other dimensional analysis methods, such as the Rayleigh method introduced in section 6.3.2, the method based on the Buckingham's Π theorems provides a better generalized strategy for obtai-

较之于其他量纲分析方法，如 6.3.2 节介绍的瑞利法，基于白金汉姆 Π 定理的量纲分析方法更具有普适性，是一种比瑞利法更形式

ning a solution. It is a formalization of the Rayleigh method of dimensional analysis and provides a strategy for reorganizing the given variables into sets of dimensionless parameters from, even though the system equation is unknown.

In the Π theorems, straightforwardly, it is stated that if a physical situation involves a certain number n of physical variables, this situation can be described by an original equation written in terms of a set of p ($p=n-r$) dimensionless parameters (Π_1, Π_2, $\cdots$, Π_p), where r is the number of fundamental dimensions. Π_1, Π_2, etc., are constructed from the original variables.

Several methods can be used to form the dimensionless parameters (i. e., Π terms). Now the method of repeating variables will be described in detail. Several steps are included in this method.

Step 1 Select a number n of variables involved in the problem under investigation. The variables include any quantity, as well as dimensional and nondimensional constants. It is suggested that the variables should include those used to describe the geometry of the system (such as diameter), to define any fluid properties (such as viscosity), and to indicate external effects that influence the system (such as driving pressure drop).

Step 2 Express each of the variables using basic dimensions. For a typical fluid mechanics problem the basic dimensions will be either M, L, and T or F, L, and T. The basic dimensions for typical variables found in fluid mechanics problems are listed in Table 1.2.

Step 3 Determine the number of fundamental dimensions, r, i. e., the number of dimensions involved in all variables.

Step 4 Select a set of repeating variables, where the number required is equal to the number of fundamental dimensions. Repeating variables are those may appear in all or most of the Π terms, and have obvious influence on the problem. Before commencing an analysis, repeating variables should be determined reasonably. The choice of repeating variables is of no strict restrictions but some rules should be followed.

① The repeating variables must cover all the dimensions (M, L, T), but that is not to say that each variable must contain M, L and T.
② A combination of the repeating variables must not form a dimensionless group.
③ The repeating variables do not have to appear in all Π groups.
④ The repeating variables selected should be measurable in experiment. For example, for a pipe flow the pipe diameter is more important and measurable than roughness even though both have the same dimension of L.

Step 5 Form a Π term by multiplying one of the non-repeating variables by the product of the repeating variables with their respective exponents. Each Π term should comprise one of the non-repeating variables, all the repeating variables and the exponents which are to be determined to make the Π term be dimensionless.

Step 6 Repeat Step 5 for each of the remaining non-repeating variables. The number of Π terms will correspond to the number of non-repeating variables.

Step 7 Check all the constructed Π terms to make sure they are dimensionless, i.e., determine all the exponents in these Π terms.

Step 8 Establish a qualitative relationship among all the Π terms. Next what we need to work with are the Π terms but not the original variables. However, it should be noted that the dimensional analysis can result in only a simple relationship but can not give the complete function for all Π terms. The actual functional relationship among the Π terms must be determined according to experimental result.

In fluid mechanics it is usually possible to select velocity, characteristic length and density or viscosity as the three repeating variables. Different repeating variables will result in different Π terms which are valid.

【EXAMPLE 6.3】 Establish a set of Π terms for the pressure drop, Δp, of incompressible flow in circular pipe. The physical parameters related to the flow process include:

① Pressure drop, Δp
② Length of pipe, L
③ Diameter of pipe, D
④ Roughness of inner surface of pipe, e
⑤ Velocity of flow, v
⑥ Density of fluid, ρ
⑦ Viscosity of fluid, μ

物理参数包括：

① 压降 Δp
② 管长 L
③ 管径 D
④ 管子表面粗糙度 e
⑤ 流速 v
⑥ 流体密度 ρ
⑦ 流体黏度 μ

SOLUTION The relationship function for the seven variables holds

解 上述变量间的关系可表示为

$$f_1(\Delta p, v, L, D, e, \rho, \mu) = 0$$

It is obvious that $n = 7$. The fundamental dimensions include M, L and T, hence $r = 3$. Then 4 $(n - r = 4)$ Π groups can be established, and

变量总数 $n = 7$，涉及的基本量纲为 M、L 和 T，故 $r = 3$，可建立的 Π 项个数为 4。于是

$$f_2(\Pi_1, \Pi_2, \Pi_3, \Pi_4) = 0$$

Taking D, v, ρ as the fundamental variables, we have

选择 D、v、ρ 为基本量纲，有

$$\Pi_1 = \Delta p D^{a_1} v^{b_1} \rho^{c_1}, \Pi_2 = L D^{a_2} v^{b_2} \rho^{c_2}, \Pi_3 = e D^{a_3} v^{b_3} \rho^{c_3}, \Pi_4 = \mu D^{a_4} v^{b_4} \rho^{c_4}$$

For each Π term, the dimensions of the left and right sides are consistent, so that

各 Π 项左右两侧的量纲一致，于是

$$\begin{cases} 0 = 1 + c_1 \\ 0 = -1 + a_1 + b_1 - 3c_1 \\ 0 = -2 - b_1 \end{cases}, \begin{cases} 0 = c_2 \\ 0 = 1 + a_2 + b_2 - 3c_2 \\ 0 = -b_2 \end{cases}, \begin{cases} 0 = c_3 \\ 0 = 1 + a_3 + b_3 - 3c_3 \\ 0 = -b_3 \end{cases}, \begin{cases} 0 = 1 + c_4 \\ 0 = -1 + a_4 + b_4 - 3c_4 \\ 0 = -1 - b_4 \end{cases}$$

Solving the four equation groups gives

求解以上 4 个方程组，得

$$\begin{cases} a_1 = 0 \\ b_1 = -2 \\ c_1 = -1 \end{cases}, \begin{cases} a_2 = -1 \\ b_2 = 0 \\ c_2 = 0 \end{cases}, \begin{cases} a_3 = -1 \\ b_3 = 0 \\ c_3 = 0 \end{cases}, \begin{cases} a_4 = -1 \\ b_4 = -1 \\ c_4 = -1 \end{cases}$$

Then the 4 terms can be expressed as

于是 4 个项为

$$\Pi_1 = \frac{\Delta p}{\rho v^2} = Eu, \Pi_2 = \frac{L}{D}, \Pi_3 = \frac{e}{D}, \Pi_4 = \frac{\mu}{Dv\rho} = \frac{1}{Re}$$

Applying the Π theorems gives

根据 Π 定理，有

$$Eu = \frac{\Delta p}{\rho v^2} = f_3\left(Re, \frac{L}{D}, \frac{e}{D}\right)$$

6.4 Similitude and Modeling

A model is used to predict the behavior of the physical system which is called the prototype. It is widely used in engineering projects involving aircraft, ships, rivers, harbors, dams, pipe transportations, and so on.

Model resembles the prototype, but their sizes may be different, and both the fluids and operating conditions may be different. A model is usually smaller than the prototype, but in some cases where prototypes are too small to perform investigations, it may have advantages to build a model which is larger than prototype. As long as the model is valid, it is possible to predict the behavior of the prototype under a certain set of conditions. In this section the modeling procedure is introduced.

6.4.1 Approximate Model of Fluid Mechanics Problem

(1) Approximate model

The similitude principle provides a theoretical basis for model investigation in fluid mechanics. To commence a model investigation, it is necessary to ensure that the flow in model is similar to that in prototype. The details to be considered are as follows.
① Flows in prototype and model should be governed by the same group of equations. This can be realized only when the medium in model is the same as that in prototype; if not, there may exist different functions that describe the relationship between their physical properties and temperature.
② The passage profile of model should be similar to that of prototype.
③ Fluid physical properties (density, viscosity, etc.) in model should be similar to that in prototype point to point.
④ The velocity distribution on the inlet surface of model is similar to that of prototype, as well as outlet.

6.4 相似与模化

模型可用来预测原型的特性。模型在工程上的应用十分广泛，如：飞行器、船舶、河流、港口、大坝、输运管道等。

模型与原型相似，但尺寸可以不同，模型中的介质与操作条件也可以有别于原型。模型尺寸一般小于原型，但当原型特别小时，也可采用放大的模型。只要模型正确，就可以在一定条件下正确预测原型的特性。本节将介绍整个模化过程。

6.4.1 流体力学问题的近似模型

(1) 近似模型

相似原理提供了进行模型研究的理论基础。在进行流体力学模型研究时，需要保证模型中的流动与原型中的流动相似。这需要从以下几个方面进行考虑。
① 模型与原型中的流动应由同一完整方程组描述。只有当模型与原型中的介质一样时，这种情况才能实现，否则它们的物理特性与温度之间的函数关系就不会完全相同。
② 模型与原型的过流通道应几何相似。
③ 模型与原型中对应点上流体的物理性质（密度、黏度等）相似。
④ 模型与原型进口、出口截面处的速度分布相似。

⑤ The initial flow conditions of model should be similar to that of prototype.
⑥ Furthermore the similarity laws of model should be equal to that of prototype one by one.

However, in most cases, it is very difficult to meet all the above requirements. For example, it is impossible to model the uneven distributions of density and viscosity of air or flue in a combustion chamber. For steady isothermal flow of incompressible viscous fluid, the Reynolds number, as well as Froude number, in model can not be respectively equal to that in prototype. This can be understood from the following analysis.

Suppose the similarity laws for prototype include Re_p and Fr_p, and for model, they are Re_m and Fr_m. The condition of $Re_p = Re_m$ can be expressed as

$$\frac{v_m l_m \rho_m}{\mu_m} = \frac{v_p l_p \rho_p}{\mu_p}$$

If $\mu_m/\rho_m = \mu_p/\rho_p$ and $l_m = l_p/10$, $v_m = 10 v_p$. But for $Fr_p = Fr_m$

$$\frac{v_p^2}{g_p l_p} = \frac{v_m^2}{g_m l_m} \text{ or } \frac{v_m}{v_p} = \sqrt{\frac{l_m}{l_p}}$$

In terms of $Fr_p = Fr_m$ it is required that the velocity in model should be 1/3.16 of that in prototype, i.e., $v_m = 1/3.16 v_p$. Obviously, the two velocity relationships are conflicting. On this condition, the model can be established only when the kinematic viscosities of the two fluids used in model and prototype hold the relationship as

$$\left(\frac{l_m}{l_p}\right)^{1.5} = \frac{\nu_m}{\nu_p}$$

That is, the kinematic viscosity of the fluid in model is 1/31.6 of that in prototype, and it is almost impossible to be realized.

⑤ 模型流动与原型流动的初始条件相似。
⑥ 模型与原型的相似准则一一相等。

但是，要完全满足上述条件是很困难的。例如在模化燃烧室内的空气、烟气流动情况时，要保证模型中密度和黏度分布与原型相似是极其困难的。对不可压缩黏性流体的稳定等温流动，很难同时保证模型与原型中的 Re 和 Fr 相等，采用下例加以说明。

假设某一原型流动的定性准则为 Re_p 和 Fr_p，相应的模型中为 Re_m 和 Fr_m。根据 $Re_p = Re_m$ 有

如果 $\mu_m/\rho_m = \mu_p/\rho_p$，且 $l_m = l_p/10$，$v_m = 10 v_p$，但对 $Fr_p = Fr_m$，有

要保证 $Fr_p = Fr_m$，则要求模型中的速度为原型的 1/3.16，即 $v_m = 1/3.16 v_p$。显然，这与雷诺数要求的 $v_m = 10 v_p$ 是矛盾的。若要使 Fr 也满足要求，则对模型和原型流体的黏度有如下要求

要求模型流体的运动黏度为原型的 1/31.6，这几乎是无法实现的。

This example shows that the choice of a fluid used in model is restrained by the model size when there are two similarity laws to be followed. If there are three or more, other physical quantities of fluids will be restricted by each other. As a result, model investigation can not be carried out favorably.

Usually it is necessary to use an approximate model to replace the real model. In approximate model, only those dominant similarity conditions are taken into account. So the choice of key similarity laws is of great importance to carry out a reasonable model investigation.

Considering the flow of air or flue gas in combustion chamber again, the cold model is usually employed to substitute the thermal state experiment, in which the isothermal cold air is used as the experimental fluid. The cold air presents a forced convective flow with pressure change, and its flow status is determined by Re but not Fr. So only Re needs to be considered while modeling, and then the difficulty in modeling is reduced greatly. Certainly the results obtained based on cold model should be revised accordingly.

(2) Requirements for approximate model

Fluid The compressible fluid can not be used to model the incompressible fluid flow. Gas can be regarded as an incompressible fluid when the relative density change $|\Delta\rho/\rho_0|$ [$\Delta\rho/\rho_0 = (\rho-\rho_0)/\rho_0$, where ρ_0 is initial density] is smaller than 5%. If the velocity of gas (or air) is no larger than 0.3 time of the local sound speed, it can be used to model liquid (such as water) flow. Up to this point, for general thermodynamic equipment, any viscous fluid can be used in model.

Flow passage Model and prototype should be similar in geometry, and this can be realized easily. As for the surface roughness, it only has effect on the near-wall

上述分析说明，当相似准则有2个时，模型的介质选择要受模型尺寸选择的限制。当相似准则有3个或更多时，流体其他物理量也要相互制约，模型研究难以进行。

为使模型研究得以进行，必须采用近似模型代替真实模型。近似模型只考虑影响流动过程的主要相似条件。因此相似准则的选择十分重要，其决定了研究的合理性。

再以锅炉炉膛中空气、烟气的流动为例，一般用冷态模型代替热态试验，采用等温冷空气作为介质。由于冷空气是有压（强迫）流动，因此决定流动状态的是Re，而不是Fr。因而模型实验时只需考虑Re准则，建模难度大大下降。当然，冷态模化试验的结果与热态情况有偏差，要进行必要的修正。

（2）对近似模型的要求

流体 可压缩流体不可以用来模化不可压缩流动。当气体相对密度变化的绝对值$|\Delta\rho/\rho_0|$小于5%时，可将其视为不可压缩流体。当气体（或空气）的流速不超过当地声速的0.3倍时，可以用来模化液体（如水）的流动。对于一般热力设备，黏性流体均可用于模化。

流道 模型与原型的流道几何形状应相似，这种条件容易实现。表面粗糙度仅对壁面附近流体的

fluid flow, hence surface feature similarity is not indispensable in terms of large flowing space. But for turbulent flow in a rough tube, the surface roughness has a considerable effect on flow resistance, so surface roughness similarity should be ensured.

Fluid physical properties The fluid physical parameters at each point of model must be similar to that of prototype, which is difficult to be realized in the case of the fluid flow with nonuniform temperature distribution. But the flow of isothermal medium (such as cold air) can possibly be used to model the flow of non-isothermal fluid, and then the results obtained should be revised necessarily.

Velocity profiles on inlet and outlet It is required that the velocity distributions on the inlet and outlet surfaces of model should be similar to that of prototype, respectively. Experimental results have shown that the velocity field of a developed pipe flow is not related to the inlet velocity distribution, in any kinds of flow channels. This phenomenon is defined as the stability of viscous flow. Therefore, the similar inlet velocity profile can be obtained by adding a geometrically similar passage before the inlet, and then the outlet velocity distribution similarity can be automatically satisfied in the geometrically similar outlet passage.

Initial conditions For a steady flow, initial conditions are not necessary to be considered.

Similarity laws Model and prototype flows should follow the same similarity laws. Just as analyzed before, the dominant similarity laws should be determined before commencing a model investigation. Should a similitude number of model be strictly equal to that of prototype? The answer will be revealed in the following example.

According to the magnitude of Reynolds number, fluid

flow status can be classified as laminar flow ($Re < Re_1$), transitional flow ($Re_1 < Re < Re_2$) and turbulent flow ($Re > Re_2$), where Re_1 and Re_2 are the first and second critical Reynolds number, respectively (Fig. 6.3). When $Re < Re_1$, for example a laminar pipe flow, the velocity distributions at different cross sections are all parabolic and not affected by Re. This characteristic is defined as "self-modeling". When $Re > Re_1$, the velocity distribution changes obviously and is affected by Re. As Re increases continuously, the influence of Re on velocity distribution is weakened. Once $Re > Re_2$, the effect of Re disappears and the flow enters the second self-modeling zone. In the second self-modeling zone, the velocity distribution remains constant again.

分为三种：层流（$Re < Re_1$）、过渡流（$Re_1 < Re < Re_2$）和湍流（$Re > Re_2$），其中 Re_1 和 Re_2 分别称为"第一临界值"和"第二临界值"，如图 6.3 所示。当 $Re < Re_1$ 时，例如圆管中的层流流动，不论流速如何，沿横截面的速度分布形状总是一轴对称的旋转抛物面。这种特性称为"自模化"。当 $Re > Re_1$，速度分布变化明显，若 Re 继续增加，它对速度分布的影响减弱。当达 $Re > Re_2$ 后，Re 的影响消失，流动进入第二自模化区，在第二自模化区中流速分布规律不再变化。

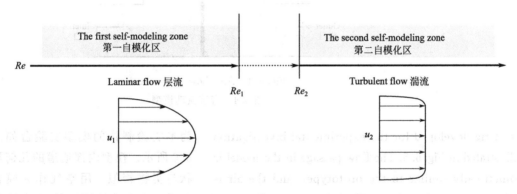

Fig. 6.3 The first and second self-modeling zones
图 6.3 第一和第二自模化区

The self-modeling characteristic introduces great convenience into modeling. So long as the flows in both model and prototype correspond to the same self-modeling zone, their velocity distributions must be similar. If Re is much larger than Re_2 in a prototype flow, Re of model is required to be just a little larger than Re_2 because on this condition, the velocity similitude for model and prototype can be ensured. Then the power of pump or fan used in the experiment can be reduced while establishing the test rig.

自模化特征给模化研究带来了极大的方便。只要原型和模型处于同一自模化区内，原型和模型中的速度分布就是相似的。当原型中的 Re 远大于第二临界值 Re_2 时，模型中的 Re 稍大于第二临界值 Re_2 即可保证速度相似。这样在建立实验装置时可减小实验中泵或风机的功率。

6.4.2 Modeling Example

To understand the modeling process for a fluid mechanics problem better, an experimental example of elbow flow is introduced in this section. The flow passage with quadrangular cross section is illustrated in Fig. 6.4, in which the flow resistance is relatively large. To reveal why there exists so large flow resistance in the passage, experiment is carried out and some measurements to reduce drag force will be proposed.

6.4.2 模化实例

为了更好地理解流体力学问题的模化过程，本节介绍一个弯道流动示例，流道截面为矩形，如图6.4所示。弯道中的流动阻力十分显著。通过实验研究，确定流动阻力产生的原因，并提出减小阻力的方法。

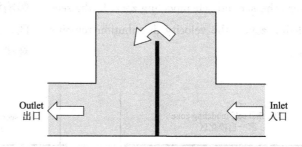

Fig. 6.4　An elbow flow model
图 6.4　弯管流动模型

A test rig developed for the experimental investigation is illustrated in Fig. 6.5. The flow passage in the model is geometrically similar to the prototype, and the air is used as the model medium. The inlet volume flow rate Q is measured by a pitot tube and the pressure difference between the inlet and outlet, Δp, is measured by a U-tube pressure gauge. The density, ρ, and viscosity, μ, of air can be determined according to its pressure and temperature.

为本实验制作的模型实验台如图6.5所示。模型内部通道的几何形状与实物相似，用空气作为模化介质。入口的体积流量Q由皮托管测量，进出口的压差Δp由U形管压差计测量。根据供风的压力及温度确定空气的密度ρ和黏度μ。

The height and width of the flow cross section are a and b respectively, so $A=ab$ (A is flow cross section area). The mean flow velocity at the inlet is

流道截面的高和宽分别为a和b，流道面积$A=ab$。平均速度为

$$v=\frac{Q}{A}=\frac{Q}{ab}$$

The equivalent diameter of the flow cross section, d_e, is

流动截面的当量直径d_e为

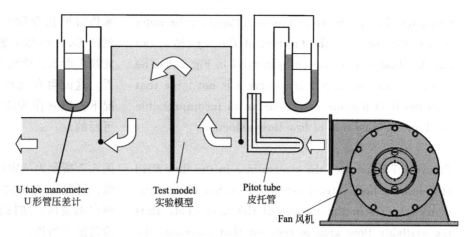

Fig. 6.5 A test rig developed for the experimental investigation of elbow flow

图 6.5 弯管流动的实验装置

$$d_e = \frac{4A}{L}$$

where L is length of wetted perimeter. So $d_e = 2ab/(a+b)$ and then

式中，L 为湿润周边的长度。于是 $d_e = 2ab/(a+b)$，且有

$$Re = \frac{v d_e \rho}{\mu} = \frac{2Q}{(a+b)\nu} \text{ and } Eu = \frac{\Delta p}{\rho v^2} = \frac{\Delta p a^2 b^2}{\rho Q^2}$$

By testing the change of Δp with the change of Q, the relationship between Re and Eu can be obtained, as graphically expressed by the curve 2 in Fig. 6.6. It can be seen that the drag coefficient of the elbow is actually large.

改变流量 Q，测出相应的 Δp，可得到 Re 和 Eu 的关系，如图 6.6 中曲线 2 所示。由图可见，弯道的阻力系数很大。

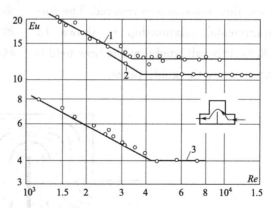

Fig. 6.6 The sketch map of relationship between Eu and Re $[Eu=f(Re)]$

图 6.6 Eu 和 Re 的关系示意图 $[Eu=f(Re)]$

1—water model 水模型；2—air model 空气模型；3—water model after adding guide vanes 加导流叶片后的水模型

Water can be used as the model medium by substituting a

把图 6.5 中的风机换成泵，则可用

pump for the fan shown in Fig. 6.5. Based on the water model, the curve for the function of $Eu = f(Re)$ can also be obtained, as shown by curve 1 in Fig. 6.6. The distance between curve 1 and curve 2 is not large that indicates it is feasible to use air as an incompressible fluid on the condition of low flow velocity.

Figure 6.7 illustrates the streamlines in the flow field at a volume flow rate. It can be seen that there exist many vortexes in the corners of the flow field, thus the available flow area is reduced that increases the flow resistance.

水作为模化介质。采用水模型得到的 $Eu=f(Re)$ 曲线如图 6.6 中曲线 1 所示。曲线 1 和曲线 2 很接近,这说明在气体流速不大的情况下,把它作为不可压缩流体是可行的。

图 6.7 所示为一定体积流量下的流线。可见,流场的拐角处出现了很多涡流区,实际流通面积减小,流动阻力增加。

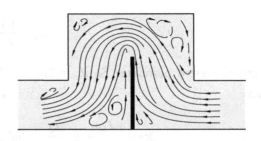

Fig. 6.7　Streamlines in the elbow flow
图 6.7　弯管流动中的流线

Based on the experimental results, we can arrange some guide vanes at the turning points in the bending passage. As shown in Fig. 6.8, the stream pattern in the improved flow passage indicates that vortexes are eliminated and the flow resistance is reduced. The corresponding experimental relationship is shown by curve 3 in Fig. 6.6, it is validated that the flow field is significantly improved by the guide vanes.

根据上述实验结果,考虑在通道拐弯处设置导流叶片。如图 6.8 所示,改进后的流道中的流线说明旋涡消失,流动阻力下降。相关实验结果如图 6.6 中的曲线 3 所示。实验证明,加导叶后,流动状况显著改善。

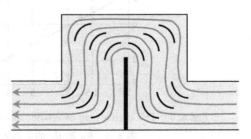

Fig. 6.8　Streamlines in elbow flow with guide vanes
图 6.8　加导流叶片后弯管流动中的流线

The effect of guide vanes is remarkable and the drag coefficient in the improved flow passage is only 1/3 of that in the original one. The horizontal sections of the curves means that Eu is constant and the flows fall in the second self-modeling zone. The critical value of Re_2 equals to 3.5×10^3 and 4.5×10^3 for the flows without and with guide vanes, respectively. That is due to the large degree of turbulence in the flow without guide vanes. In engineering, most of the flows fall in the second self-modeling zone.

Exercises

6.1 The Reynolds number, Re, is a very important parameter in fluid mechanics. Verify that the Reynolds number is dimensionless, using the MLT system for basic dimensions, and determine its value for water flowing at a velocity of 2m/s through a 20-cm-diameter pipe.

6.2 What are the dimensions of density, pressure, specific weight, surface tension, and dynamic viscosity in the MLT system?

6.3 At a sudden contraction in a pipe the diameter changes from D_1 to D_2. The pressure drop, Δp, which develops across the contraction is a function of the velocity, v, in the larger pipe, the fluid density, ρ, and viscosity, μ. Use as repeating variables to determine a suitable set of dimensionless parameters. Why would it be incorrect to include the velocity in the smaller pipe as an additional variable?

6.4 The pressure rise Δp across a pump is related to the impeller diameter, D, the fluid density, ρ, the rotational speed, ω, and the flow rate, Q. Determine a suitable set of dimensionless parameters for the flow. Δp can be expressed as

$$\Delta p = f(D, \rho, \omega, Q)$$

6.5 The pressure drop, Δp, along a straight pipe with diameter, D, has been experimentally studied, and it is observed that for laminar flow of a given fluid and pipe, the pressure drop varies directly with the distance, l, between pressure taps. Assume that Δp is a function of D, l, the velocity, v, and the fluid viscosity, μ, Use dimensional analysis to deduce how the pressure drop varies with pipe diameter.

6.6 The drag force exerting on a ball is 5N when it is emerged in water flows at the velocity of 2m/s. Another ball which is twice larger is located in wind tunnel. On the condition of dynamic similarity, determine the wind velocity and find the drag force on the larger ball when $\nu_{air}/\nu_{water}=13$ and $\rho_{air}=1.28\text{kg/m}^3$.

6.7 A ball with the diameter of d and density of ρ drops in a fluid with the density, ρ_1, and viscosity, μ_1. Determine the sedimentation velocity of the ball, v, using dimensional analysis method.

6.8 An experiment is carried out in a wind tunnel for the automobile with the real running speed of 30m/s. If the experimental wind speed is 45m/s, find the geometrical ratio of prototype to model. If the drag force on the model is 2kN, find the drag force on the prototype.

6.9 The air is used to substitute for water to perform an experiment to test the local drag coefficient of a valve at the temperature of 20℃. If the water velocity is 2.5m/s, determine the air velocity in the experiment. Furthermore, what is the ratio of the pressure head losses arising from air and water?

6.10 Fig. 6.9 illustrates an orifice plate flowmeter for testing air flow rate, where $d=100$mm and $D=200$mm. Now we use water to verify this flowmeter. The minimum flow rate Q_m is 0.02m³/s when the discharge coefficient remains constant and the mercury pressure gauge reading (between points 1 and 2) is 50mm. On

6.5 实验研究发现，直径为 D 的直管内层流流动的压降 Δp 与两个压力测点之间的距离 l 有关。令 Δp 为 D、l、速度 v 和黏度 μ 的函数。采用量纲分析法确定压降与管径之间的关系。

6.6 放在流速为 2m/s 水中的圆球受到的阻力为 5N。另一个直径为其两倍的圆球置于一风洞中。求在动力相似条件下风速的大小及球所受的阻力。已知 $\nu_{air}/\nu_{water}=13$，$\rho_{air}=1.28\text{kg/m}^3$。

6.7 有一直径为 d，密度为 ρ 的圆球在充满密度为 ρ_1、黏度为 μ_1 的无限空间中沉降。试用量纲分析法确定其沉降速度 v 的关系式。

6.8 汽车行驶速度为 30m/s，拟在风洞中进行模化实验。实验风速为 45m/s，求原型与模型的长度尺度比。又问如模型上测得的阻力为 2kN，求原型汽车所受的阻力。

6.9 利用空气测定阀门的局部阻力系数，试验温度为 20℃。水速为 2.5m/s，问空气速度应为多少？空气和水在经过阀门后的压力损失的比值应为多少？

6.10 图 6.9 所示为空气孔板流量计，$d=100$mm，$D=200$mm。用水校验该测量计。孔板流量系数不变时的最小流量为 $Q_m=0.02$m³/s，水银压差计读数为 $h=50$mm。已知流量系数不变的最小

the condition that the discharge coefficient is constant, the Reynolds number that corresponds to the minimum flow rate is the second critical Reynolds number. Calculate: (a) the minimum flow rate when the flowmeter is used to test air flow rate; (b) the mercury pressure gauge reading under this air flow rate. The test temperature is 20℃.

流量对应的雷诺数是第二临界雷诺数。
计算：(a) 当孔板用来测定空气流量时，最小流量是多少？(b) 该流量下的水银压差计读数是多少？试验温度为 20℃。

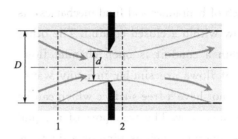

Fig. 6.9　Figure for exercise 6.10
图 6.9　习题 6.10 图

Chapter 7 Pipe Flow
第7章 管内流动

Pipe flow, a branch of hydraulics and fluid mechanics, is a type of fluid flow within a closed conduit. The other type of flow within a conduit is open channel flow. These two types of flows are similar in many ways, but pipe flow does not have a free surface which can be found in open-channel flow. The transport of a liquid or gas in a closed conduit, such as pipe or a duct, is extremely important in our daily operations and also there is a wide variety of applications of pipe flow in engineering (Fig. 7.1).

流体在封闭通道的流动称为管内流动，是水力学和流体力学的一个分支；此外，开口渠流动也属管内流动。这两种流动在很多方面是相似的，但敞口流动有自由表面。液体和气体在封闭管道内的输运，如管内流动，在我们的日常生活中起到了十分重要的作用，在工程上也有大量的应用，如图7.1所示。

Fig. 7.1 Pipe transportation
图 7.1 管道输送

A typical pipe flow system with basic components is shown in Fig. 7.2. It may include the pipes, various fittings used to connect the individual pipes, the valves to control flow rate, and the pumps or turbines that increase or reduce energy from the fluid.

图 7.2 所示为一个典型的管路系统，基本部件包括管道、用于连接各个管路的接头、用于控制流量的阀门以及用来提供或吸收能量的泵或透平。

Even the most simple pipe systems are actually quite complex in terms of rigorous flow characteristics when performing analytical investigation. In engineering the design of a pipe system mainly concerns the layout of pipes desired, the analysis of head loss throughout the

即便对于最简单的管路系统，采用解析法研究其中的流动特征仍是复杂的。工程上，管路系统的设计包括管路的布局、管路阻力损失的分析以及对高温高压工况

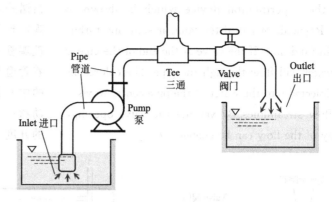

Fig. 7.2 A typical pipe system
图 7.2 典型的管路系统

pipes and the calculation of stress distribution in the system when it is working at elevated temperature and high pressure.

In this chapter we will analyze some pipe flow topics, such as laminar and turbulent flows in constant diameter pipes as well as the major and minor pressure head loss in pipeline. By a combination of experimental data with theoretical considerations, fluid viscosity and surface roughness effects are considered on the flow resistance characteristics. Finally the head loss calculation methods for single, series and parallel pipelines will be introduced.

7.1 General Characteristics of Pipe Flow

The behavior of pipe flow is affected mainly by viscous force, gravity force and inertial force on fluid particles. Depending on the ratio of viscous force to inertial force, as characterized by the Reynolds number, the flow can be either laminar or turbulent.

In fluid mechanics, the Reynolds number (Re) is a dimensionless quantity which is used to predict flow status in different situations. The concept was introduced by George Gabriel Stokes in 1851, but the Reynolds number was named after Osborne Reynolds, who popularized its use in 1883.

下管系应力分布的计算等。

本章首先介绍管内流动的一些基本概念，包括层流、湍流、主要和次要阻力损失等。结合实验数据与理论分析，研究流体黏度和表面粗糙度对流动阻力特性的影响。最后介绍常见的简单管路、串联和并联管路中的流动阻力损失的计算方法。

7.1 管内流动的一般特性

管内流动的特性与流体质点的黏性力、重力以及惯性力有关。根据黏性力和惯性力比值，即雷诺数，流动可分为层流和湍流。

流体力学中，雷诺数 Re 是一个无量纲数，可用来判断流体的流动状态。这一概念由 Stokes 于 1851 年提出，Reynolds 在 1883 年开始推广应用。

Based on the experimental device which is shown in Fig. 7.3, Reynolds observed the laminar state and turbulent state in fluid flow. If the valve at the end of the pipe is opened, water will flow through the pipe. The colored filament is injected into the water at the pipe center to visualize the flow streamlines. By varying the valve opening, the velocity of the flow can be changed.

雷诺采用如图 7.3 所示的实验方法观察到了层流和湍流现象。打开玻璃管末端的节流阀后,水将沿着管道流动。将有色流线从管路的中心位置注入水中,以观察流动状态。实验中通过调节节流阀的开度,可以改变水流速度。

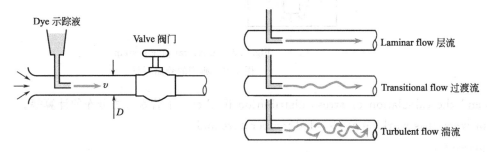

Fig. 7.3 Pipe flow status observed from the Reynolds experiment
图 7.3 由雷诺实验观察到的管内流动状态

The colored filament is straight and smooth for low speeds, it does not mix with the surrounding fluid and there is no macro transverse of the particles. The flow at this state is a laminar flow. As the velocity increases, the colored filament starts to fluctuate and the transverse motion of the particles becomes outstanding. Then the flow is not laminar any more. However, the colored filament breaks off and disperses almost uniformly for high velocities, and this state is termed as a turbulent flow.

流速较低时,有色流线呈一直线且光滑,与周围流体互不相混,质点没有宏观的横向运动,这种流动状态称为层流;当水流速度增大时,有色流线开始呈波纹状,流体质点有明显的横向运动,流动不再是层流。当流速较高时,有色流线断裂并均匀扩散,此时流动状态称为湍流。

Reynolds also found that there is a critical velocity, v_{cr}, from laminar state to turbulent state. v_{cr} depends on the kinematic viscosity of the fluid, ν ($\nu = \mu/\rho$), and the diameter of pipe, d. The Reynolds number, Re, is just a dimensionless group of v, ν and d.

Reynolds 发现在层流和湍流之间存在一个临界速度 v_{cr}。v_{cr} 与流体的运动黏度 ν ($\nu=\mu/\rho$) 和管径 d 有关。雷诺数 Re 就是 v、ν 和 d 这三个参数的无量纲组合。

$$Re = \frac{\rho v d}{\mu} = \frac{\rho v^2}{\mu v/d} = \frac{\text{Inertia force 惯性力}}{\text{Viscous force 黏性力}} \quad (7.1)$$

As is introduced in chapter 6, the Reynolds number is an evaluation of the ratio of inertial force to viscous force and consequently quantifies the relative importance of these two types of forces for given flow conditions.

正如第 6 章介绍的,雷诺数是流体所受惯性力和黏性力的比值,可以用来衡量这两种力对流动影响的重要程度。

When Re is small, the viscous force is dominant and can weaken the disturbance that initiates the chaos of particles, so the flow maintains laminar state. When Re is large, the flow is dominated by inertial forces which tend to produce chaotic eddies, vortices and other flow instabilities, and then the flow becomes turbulent.

The transition from laminar to turbulent state is related to the Reynolds number. There exist two critical Reynolds numbers, Re_{cr1} (2300) and Re_{cr2} (4000), to distinguish the three states, such as laminar state, transitional state and turbulent state.

Under laminar state ($Re \leqslant Re_{cr1}$), the flow is termed the steady laminar flow. Even if the flow is disturbed, it can return to be laminar once the disturbance is removed.

Under turbulent state ($Re > Re_{cr2}$), even a tiny disturbance can make the flow unstable, so it is an unsteady state. In turbulent flow, unsteady vortices appear on many scales and interact with each other. The drag due to boundary layer skin friction increases, and the boundary layer structure and the location of boundary layer separation change, sometimes resulting in a reduce of overall drag.

The transitional state falls in between laminar state and turbulent state ($Re_{cr1} < Re \leqslant Re_{cr2}$). Under this state, the flow is uncontrollable. Although laminar-turbulent transition is not governed by the Reynolds number, the transitional situation will occur if the object size is increased, or the viscosity of the fluid is decreased, or the density of the fluid is increased.

It should be indicated that two critical Reynolds numbers, Re_{cr1} and Re_{cr2}, are not unchangeable and may be affected by the entrance condition of the pipe and the external disturbance.

Re 较小,说明黏性作用力较大,它能削弱和消除引起流体质点乱运动的扰动,使流动保持层流状态;Re 较大时,惯性力处于支配地位,惯性力会产生旋涡和流动不稳定,流动进入湍流状态。

雷诺实验中所发生的流动状态变化与雷诺数有关。根据两个临界雷诺数,即 Re_{cr1}（2300）和 Re_{cr2}（4000）,可区别层流、过渡流和湍流这3种流动状态。

当 $Re \leqslant Re_{cr1}$ 时,流动称为稳定层流。即便有外界扰动,一旦扰动消失,流动仍能恢复到层流状态。

当 $Re > Re_{cr2}$ 时,任何微扰动都能使流动失稳,流动处于不稳定状态。湍流中会出现各种尺度的、相互作用的旋涡。边界层的摩擦阻力增加,边界层结构和分离位置改变,有时会导致总体流动阻力的减小。

当 $Re_{cr1} < Re \leqslant Re_{cr2}$ 时,流动介于层流和湍流之间,称为过渡流,过渡流不可控。虽然层流-湍流的过渡不受雷诺数控制,但通过增大物体的尺寸、减小黏度或增大密度均可获得同样的过渡状态。

临界雷诺数 Re_{cr1}、Re_{cr2} 并非固定不变,它们与管道入口状况及外界扰动情况有关。

7.2 Laminar Flow in Circular Pipe

The flows in circular pipe and circular jacket can be frequently seen in engineering applications. The ratio of length to radius of common pipes is usually much larger than 1 so that effect of the inlet and outlet can be ignored. The flow can be regarded as a fully developed one-dimensional flow. Usually a pipe flow is caused by the pressure difference between the inlet and outlet and if the pipe is not horizontal, gravity effect on the flow should be considered.

Fig. 7.4 shows a declining pipe flow model in the cylindrical coordinate system with the radial coordinate of r and the axial coordinate of z. Since it is a developed one-dimensional flow, the velocity is the function of r, i.e., $v_z = v_z(r)$. p_i and p_o are the inlet and outlet pressures, respectively, and β is the separation angle between the z axis and the force of gravity.

7.2 圆管中的层流

圆管和圆形套管内的流动是工程实际中最常见的流动形式。由于一般管路的管长与管径之比远大于1，所以进出口区的影响可忽略，管内流动为充分发展的一维流动。流动由进出口两端的压差产生，对于非水平管道，还要考虑重力对流动的影响。

如图7.4所示的圆管内流体，在柱坐标系中，r为径向坐标，z为轴向坐标。由于是一维充分发展流动，管内流体的速度v_z应是r的函数，即$v_z = v_z(r)$。p_i和p_o分别为进出口压力，β为管子轴线与重力方向的夹角。

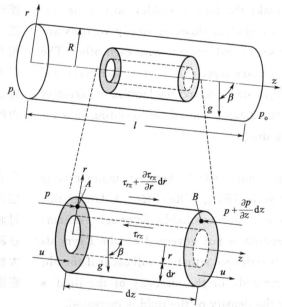

Fig. 7.4 Pipe flow model and the loads on representative element
图7.4 管内流动模型及作用于体元的力

An annular representative element is extracted from the model and all the stresses that have effects of the flow are denoted on the element.

从流体模型中取出一个环形代表体元，并将所有的应力在体元上标出。

The length of the representative element is dz, the inner diameter is r and the wall thickness is dr. p and $p + \frac{\partial p}{\partial z} dz$ are the pressures on the upstream and downstream surfaces, respectively, and τ_{rz} and $\tau_{rz} + \frac{\partial \tau_{rz}}{\partial r} dr$ are the shear stresses on the internal and external cylindrical surfaces, respectively.

代表体元的长度为 dz，内径为 r，环厚为 dr。p 和 $p + \frac{\partial p}{\partial z} dz$ 分别为体元上游和下游的压力，内外侧圆柱表面的切应力分别为 τ_{rz} 和 $\tau_{rz} + \frac{\partial \tau_{rz}}{\partial r} dr$。

Since it is a steady one-dimensional incompressible flow, the velocity gradient along the z axis is zero, i.e., $\frac{\partial v_z}{\partial z} = 0$. Moreover, the pipe is of uniform cross-section, so the average velocities at the inlet and outlet are the same, and that results in the same momentum flow rate of $\rho v^2 \times 2\pi r dr$.

对一维不可压缩稳态流动，速度沿 z 方向的梯度为 0，即 $\frac{\partial v_z}{\partial z} = 0$。又由于进出口截面面积相等，所以输入输出微元体的速度相同，故进出口截面动量流量均为 $\rho v^2 \times 2\pi r dr$。

The forces acting on the representative element include the shearing force, gravity force and pressure force.

作用于代表体元上的力包括切应力、重力和压力。

The total shearing force on the surfaces, $F_{\tau,z}$, holds

切应力的总作用力 $F_{\tau,z}$ 为

$$F_{\tau,z} = -\tau_{rz} \times 2\pi r dz + \left(\tau_{rz} + \frac{\partial \tau_{rz}}{\partial r} dr\right) \times 2\pi(r+dr) dz = \frac{\partial \tau_{rz}}{\partial r}(r+dr) \times 2\pi dr dz + \tau_{rz} \times 2\pi dr dz$$

The axial component of gravity force, $F_{g,z}$, is

重力的轴向分量 $F_{g,z}$ 为

$$F_{g,z} = \rho g \cos\beta \times 2\pi r dr dz$$

Due to the effect of gravity force, the pressure distribution is not axisymmetric about the pipe axis. But it can be validated that for a developed flow, the pressure difference between two corresponding points A and B (the two points have the same coordinates of both r and θ) on any two end faces with the distance of dz is constant and is independent of r and θ. The pressure difference is

由于重力的作用，压力分布不具有轴对称性。但可以证明，对于充分发展的流动，相距 dz 的任意两端面上的两个对应点 A 和 B（r、θ 相同）的压力差是定值，与 r 和 θ 无关。压力差为

$$p + \frac{\partial p}{\partial z} dz - p = \frac{\partial p}{\partial z} dz$$

So the resultant pressure force exerting on the two end faces of the representative element can be calculated by

因此体元上下游端面上压力的总作用力为

$$F_{p,z} = -\left(\frac{\partial p}{\partial z}dz\right) \times 2\pi r\,dr = -\frac{\partial p}{\partial z} \times 2\pi r\,dr\,dz$$

Summation of all the three forces gives the resultant force along the z direction as

体元在 z 方向所受总力为

$$\sum F_z = F_{\tau,z} + F_{g,z} + F_{p,z}$$

By omitting the high-order terms, the simplified equation is obtained

略去高阶微量后，简化后的方程为

$$\sum F_z = \left[\frac{1}{r}\frac{\partial(r\tau_{rz})}{\partial r} - \frac{\partial p}{\partial z} + \rho g\cos\beta\right] \times 2\pi r\,dr\,dz$$

According to the momentum conservation principle, the net momentum flow rate is zero, so $\sum F_z = 0$. Thereby

根据动量守恒原理，进出口净动量流量为零，所以 $\sum F_z = 0$。于是

$$\frac{1}{r}\frac{\partial(r\tau_{rz})}{\partial r} = \frac{\partial p}{\partial z} - \rho g\cos\beta = \frac{\partial p^*}{\partial z} \quad (7.2)$$

where $p^* = p - \rho g z\cos\beta$.

式中，$p^* = p - \rho g z\cos\beta$。

It can be verified that the pressure gradient on any cross-section, $\partial p^*/\partial z$, is independent of r or θ, while τ_{rz} is the function of r only. Therefore both of the two sides of Eq. (7.2) should be equal to a constant. so

可以证明，同一端面上压力的压力梯度 $\partial p^*/\partial z$ 与 r、θ 无关，而 τ_{rz} 只是 r 的函数。故式(7.2)两边必为同一常数。于是

$$\frac{\partial p^*}{\partial z} = \text{constant} = -\frac{\Delta p^*}{l}$$

$$\Delta p^* = p_i - p_o + \rho g l\cos\beta$$

Then τ_{rz} can be expressed as

τ_{rz} 可表示为

$$\tau_{rz} = -\frac{\Delta p^*}{l}\frac{r}{2} + \frac{C_1}{r} \quad (7.3)$$

According to the Newton's shear law

根据牛顿内摩擦定律

$$\tau_{rz} = \mu\left(\frac{du}{dr}\right)$$

the differential governing equation for the pipe flow is

管内流动的微分方程为

$$\frac{du}{dr} = -\frac{\Delta p^*}{l}\frac{r}{2\mu} + \frac{C_1}{r\mu} \quad (7.4)$$

Integration of Eq. (7.4) gives

式(7.4) 积分得

$$u = -\frac{\Delta p^*}{l}\frac{r^2}{4\mu} + \frac{C_1}{\mu}\ln r + C_2 \quad (7.5)$$

where the two constants C_1 and C_2 can be solved using the two velocity boundary conditions at the wall boundary and centerline

式中的待定常数 C_1 和 C_2 可根据管道壁面和中心处的速度边界条件求解

$$\left.\frac{du}{dr}\right|_{r=0} = 0, \quad u|_{r=R} = 0$$

Thereby C_1 and C_2 are

C_1 和 C_2 为

$$C_1 = 0, \quad C_2 = \frac{\Delta p^*}{l}\frac{R^2}{4\mu}$$

The shear stress is

切应力为

$$\tau_{rz} = -\frac{\Delta p^*}{l}\frac{r}{2} \quad (7.6)$$

and the velocity function yields

速度方程为

$$u = \frac{\Delta p^*}{l}\frac{R^2}{4\mu}\left(1 - \frac{r^2}{R^2}\right) \quad (7.7)$$

The maximum velocity occurs at $r=0$, and it is

最大速度位于 $r=0$ 处，即

$$v_{\max} = \frac{1}{4\mu}\frac{\Delta p^*}{l}R^2 \quad (7.8)$$

In engineering we usually use the uniform velocity, $\bar{v}$, is

平均速度 $\bar{v}$ 为

$$\bar{v} = \frac{1}{A}\int_A v_x dA = \frac{1}{\pi R^2}\int_0^R \frac{1}{4\mu}\frac{\Delta p^*}{l}(R^2 - r^2)2\pi r dr = \frac{1}{\pi R^2}\frac{\pi R^4 \Delta p^*}{8\mu l} = \frac{\Delta p^* R^2}{8\mu l} \quad (7.9)$$

It can be seen that the average velocity is a half of the maximum velocity.

可见，层流时管内平均流速为管中心最大流速之半。

The volume flow rate can be calculated by

体积流量为

$$Q = \bar{v}A = \frac{\pi R^4 \Delta p^*}{8\mu l} \quad (7.10)$$

The viscosity of a fluid can be tested according to Eq. (7.10). For a developed laminar flow in the pipe with the specified diameter and length, the viscosity can be calculated once the pressure drop through the pipe is measured.

利用流量公式 (7.10)，可以测定流体的黏度。对给定直径和长度的管道内的充分发展的层流，只要测出压差，就可以得到流体的黏度。

In chapter 5, this pipe flow is analyzed by directly using the Navier-Stokes equations without performing the force analysis for an infinitesimal element. It is hoped that the procedure introduced in this section may help you understand the differential analysis in the cylindrical coordinate system.

对于该管内层流流动问题，第 5 章直接采用 Navier-Stokes 方程进行求解，而未进行受力分析。这里介绍的方法能够帮助读者理解圆柱坐标系下的微分分析方法。

7.3 Turbulent Flow in Circular Pipe

7.3 圆管中的湍流

Turbulent pipe flow is actually more likely to occur than laminar flow in engineering. Thus it is necessary to obtain similar flow governing equation for turbulent pipe flow. However, turbulent flow is much more complicated than laminar flow, because it contains low momentum diffusion, high momentum convection, rapid variation of pressure and the temporal and spatial velocity change. Due to the elusive motion of mass particles in turbulence, most of the previous attempts in exploiting the velocity distribution are semi-empirical or theoretical based on certain assumptions.

工程上，管内流动多以湍流的形式出现，像研究层流那样建立湍流的控制方程很有必要。然而湍流要比层流复杂得多，要考虑低动量扩散、高动量对流、压力的快速变化以及速度随空间和时间的变化。由于流体质点的运动杂乱无章，前人关于湍流速度分布的研究只是在某些假设的基础上，通过半经验、半理论的方法得到。

(1) Time-averaged physical quantities in turbulent flow

(1) 湍流中的时均物理量

Turbulent flow is usually unstable. There exist not only the velocity fluctuation but also the pressure fluctuation in a turbulent pipe flow. Although the motion of particles is governed by the motion law for viscous fluids, the differential equations for fluid motion can not be solved due to the pulsations of velocity and pressure. The statistical method is usually applied and the time-averaged values are used to substitute for the instantaneous values.

湍流是一种不稳定流动。管内湍流不但有速度脉动，还有压力脉动。虽在流动瞬间流体仍服从黏性流体的运动规律，但由于脉动的存在使得运动微分方程无法求解。一般采用统计方法，并用时均值代替瞬时值。

Turbulent motion can be understood as that the fluid particles move at the time-averaged velocity and pressure. If the time-averaged parameters are constant, the flow is steady and in most engineering pipes, the flow can be regarded to be steady so that the laws for steady flow, such as the Bernoulli equation, can be applied.

湍流运动可理解为流体按时均速度和时均压力在运动。若时均参数为常数，流动是稳定的。工程管道内的湍流一般都是稳定的，故有关稳定流动的定律，如伯努利方程等均可适用。

Based on the time-averaged assumption, the pulsation effect on the fluid flow is neglected. But when analyzing the nature of turbulence, the pulsation effect should be considered; otherwise large error will be introduced in the results. For example, as we investigate turbulent flow resistance, we should consider the pulsation of particles but can not directly substitute the time-averaged velocity into the Newton's shear law.

(2) Momentum transfer in turbulent flow and additional turbulent stress

In turbulent flow, fluid particles exhibit additional transverse motion which enhances the rate of energy and momentum exchange between them thus increasing the heat transfer and the friction coefficient. The transverse momentum transfer always accelerates the particles at lower velocity while decelerates those at higher velocity. The pulsation of particles introduces a shear stress (the Reynolds stress) into fluid layers. This shear stress is different from viscous shear stress between adjacent fluid layers. Therefore, the resultant shear stress is the summation of viscous shear stress and the additional pulsation shear stress. But due to the complexity of turbulence, the Reynolds shear stress can not be determined accurately, and its solution is usually based on some assumptions.

(3) Prandtl's mixing length model

Prandtl introduced the additional concept of the mixing length, along with the idea of a boundary layer. Fig. 7.5 illustrates a turbulent flow model in the x-y coordinate system, where the x axis locates at wall boundary and the y axis is along the normal direction of the wall. Prandtl assumed that when fluid particles pulsate transversely, they have a free path, l, before they exchange momentum with the particles in adjacent layers. This free path is the mixing length which is a statistical value of all the free paths.

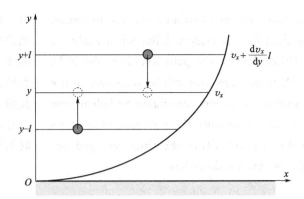

Fig. 7.5 Momentum transfer of particles in different layers
图 7.5 不同层间流体质点的动量传递模型

For wall-bounded turbulent flows, the eddy viscosity will vary with the distance from the wall, and it is the addition of the concept of a "mixing length". In the simplest wall-bounded flow model, the eddy viscosity (it is also termed the turbulent viscosity coefficient) is given by the equation

对于固壁边界的湍流，流体的涡流黏度与涡和壁面的距离有关，这个距离也可以认为是混合长度。涡流黏度（也可以定义为湍流黏性系数）可以定义为

$$\mu_t = \rho l^2 \left| \frac{dv_x}{dy} \right| \quad (7.11)$$

where $\frac{dv_x}{dy}$ is the derivative of the streamwise velocity (v_x) with respect to the wall normal direction (y direction), and l is the mixing length.

式中，$\frac{dv_x}{dy}$ 为流向速度 v_x 在壁面法线方向（y 方向）的导数；l 为混合长度。

Based on the Prandtl's mixing-length definition, the total shear stress of particles in turbulent flow can be expressed as

根据普朗特混合长度定义，湍流中流体质点的总切应力为

$$\tau = (\mu + \mu_t) \frac{dv_x}{dy} \quad (7.12)$$

(4) Velocity distribution in pipe turbulence

（4）管内湍流的速度分布

Although Eq. (7.12) gives the shear stress expression in turbulence, the velocity distribution is still impossible to be solved because, on one hand, the mixing length l is unknown and on the other hand the flow mechanism in laminar sub-layer is much different from that in bulk region. More assumptions should be proposed for the establishment of velocity equation.

虽然得出了湍流中的总切应力的表达[式(7.12)]，但仍无法求出管内湍流的速度分布函数。一是因为混合长度 l 不确定；二是因为层流底层与其外主流的流动差别很大。建立速度方程需要更多假设。

For simplicity, the laminar sub-layer velocity equation and the bulk flow velocity equation are directly given as Eq. (7.13) and Eq. (7.14), respectively.

$$\frac{v_x}{\sqrt{\tau/\rho}} = \frac{y\sqrt{\tau/\rho}}{\nu} \qquad (7.13)$$

$$\frac{v_x}{\sqrt{\tau/\rho}} = \frac{1}{K}\ln\frac{y\sqrt{\tau/\rho}}{\nu} + C \qquad (7.14)$$

where $C = \sqrt{Re} - \frac{1}{K}\ln\sqrt{Re}$. Since the term $\sqrt{\tau/\rho}$ is of the same dimensions with velocity, the variable v^* is used to denote $\sqrt{\tau/\rho}$ in the equation, thereby

$$\frac{v_x}{v^*} = \frac{yv^*}{\nu} \qquad (7.15)$$

$$\frac{v_x}{v^*} = \frac{1}{K}\ln\frac{yv^*}{\nu} + C \qquad (7.16)$$

Based on a large amount of experiments, Nikuradse determined the two constants, i.e., K and C in Eq. (7.16)

$$\frac{v_x}{v^*} = 2.5\ln\frac{yv^*}{\nu} + 5.5 = 5.75\lg\frac{yv^*}{\nu} + 5.5 \qquad (7.17)$$

Equation (7.17) is approximately valid for turbulent region, but not valid for laminar sub-layer. Figure 7.6 illustrates the relationship between v_x/v^* and yv^*/ν obtained by experimental investigation, where curve 2 corresponds to experimental result and it is almost straight when $yv^*/\nu > 50$; curve 3 is a graphical expression of Eq. (7.17) and it fits curve 2 very well when $yv^*/\nu > 50$; curve 1 is resulted from Eq. (7.15).

A simpler and exponential velocity equation for turbulent pipe flow was obtained from the experimental results, i.e.

$$\frac{v_x}{v^*} = 8.7\left(\frac{yv^*}{\nu}\right)^{1/7} \qquad (7.18)$$

This equation is appropriate when $Re < 10^5$. In laminar pipe flow, the average velocity is a half of the maximum

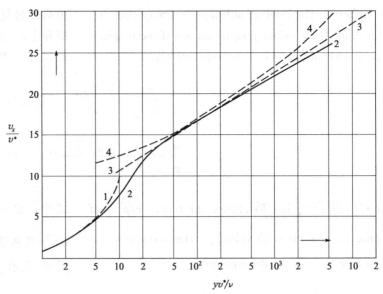

Fig. 7.6 Experimental result of turbulent pipe flow

图 7.6 管内湍流的实验结果

velocity, while in turbulent flow, the average velocity is much larger.

管内湍流的平均速度则要大得多。

Calculating the average velocity according to Eq. (7.18) gives

根据式(7.18)计算湍流的平均速度为

$$\frac{\overline{v}_x}{v^*} = \frac{1}{A}\int_A \frac{v_x}{\sqrt{\tau/\rho}} dA = \frac{1}{\pi R^2}\int_0^R 8.7\left(\frac{yv^*}{\nu}\right)^{1/7} 2\pi(R-y)dy \approx 0.82 \times 8.7\left(\frac{Rv^*}{\nu}\right)^{1/7}$$

and 且

$$\frac{v_{x\max}}{v^*} = 8.7\left(\frac{Rv^*}{\nu}\right)^{1/7}$$

so 所以有

$$\frac{\overline{v}_x}{v_{x\max}} = 0.82 \quad (7.19)$$

7.4 Pressure Head Losses in Circular Pipe

7.4 圆管中的压头损失

7.4.1 Mechanism of Flow Resistance

7.4.1 流动阻力产生的机理

(1) Generation of laminar flow resistance

(1) 层流流动阻力产生的机理

The laminar flow resistance relates to the viscosity of

层流流动阻力与流体的黏性有关。

fluid. As shown in Fig. 7.7, a Newtonian fluid is pumped into a circular glass pipe with a socket and there exists an entrance region for the fluid in terms of the velocity profile. The velocity is uniform at the inlet (A), but due to the sticking effect of pipe wall, the velocities of the molecules near the wall trend to be zero. In laminar flow, there is no large-scale transverse transportation of molecules in the fluid, but the microscopic molecular motion and Brownian motion of molecule groups still exist. The momentum exchange among the molecules with different velocities extends the wall effect from wall boundary to pipe center gradually. The momentum exchange is always accompanied by the energy consumption and that is why the frictional resistance occurs. Frictional resistance results in the reduce of average velocity, as can be seen from the velocity profiles at cross sections A and B.

如图 7.7 所示，将一种牛顿流体从圆形玻璃管喇叭口处泵入，就流体速度分布而言，存在一个入口段。流体在入口截面 A 处流速均匀分布。由于管壁的黏滞效应，紧贴管壁的流体质点的流速趋向于零。层流时没有大范围的横向质点迁移，但流体分子的微观运动和分子团的布朗运动依然存在。不同速度分子的动量交换，使壁面的影响逐步由壁面边界向管中心扩展。动量交换必然造成能量损失，这就是流体摩擦阻力产生的原因。摩擦阻力导致平均速度下降，这可从 A、B 截面上的速度分布看出。

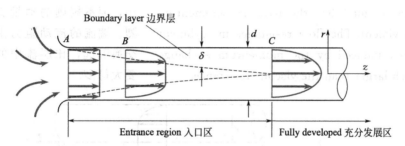

Fig. 7.7 Development of laminar flow in circular pipe
图 7.7 圆管内层流流动的发展

The flow of near-wall fluid depends on the energy that arises from the molecules or molecular groups with higher velocity. After the flow fully develops, the velocity distribution remains constant as shown on section C.

靠近壁面的流体维持流动的能量来自于更高速度的分子和分子团。流动充分发展后，形成 C 截面处所示的速度分布且保持不变。

(2) Generation of turbulent flow resistance

（2）湍流流动阻力产生的机理

Turbulent flows are always highly irregular. Under turbulent flow condition, fluid particles always move confusedly and their velocities vary with time, as shown in Fig. 7.8, and both the transverse and longitudinal velocity components are remarkable.

湍流是高度不规则的。在湍流条件下，流体质点的运动非常紊乱，其速度随时间改变，如图 7.8 所示，除了有纵向运动的速度外，还有较大的横向速度。

Chapter 7 Pipe Flow   第 7 章 管内流动   167

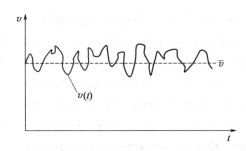

Fig. 7.8 Velocity variaiton of a particle in turbulent flow with time
图 7.8 湍流中一质点的速度随时间的变化

In a turbulent flow, there exists large-scale transverse momentum exchange of fluid particles, so that the velocity distribution is much different from that in laminar flow. Besides the motions of fluid molecules and groups, the transportations of a large number of small vortexes result in the more uniform velocity distribution in the bulk region of the field, as shown in Fig. 7.9. Since the velocity distribution is uniform, the viscous friction force is small but the irregular momentum transition is violent. The flow resistance in turbulent flow is mainly induced by the momentum exchange and it is much larger than the viscous resistance.

由于管内流体质点较大范围的横向迁移,造成湍流的速度分布及流动阻力与层流相差很大。湍流中不仅有流体分子和分子团的迁移,更主要的是有大量小旋涡的迁移,使得管内主流的速度更加均匀,如图 7.9 所示。湍流中速度分布较均匀,因而黏性摩擦力很小,但不规则的动量交换非常剧烈。湍流的流动阻力主要由动量交换引起,比层流中的黏性阻力要大得多。

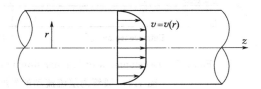

Fig. 7.9 Velocity profile in turbulent pipe flow
图 7.9 管内湍流的速度分布

Turbulent boundary layer is different from laminar boundary layer. Similar to that in laminar flow, the thin fluid layer sticking to the wall boundary in turbulent flow is also static. Due to viscous force, this layer has a drag effect on the adjacent fluid layer. On the inlet cross-section, the turbulence or boundary layer does not fully develop, thus the boundary layer is very thin and can be classified as laminar boundary layer. When it flows apart from the inlet, the turbulence develops continuously and the turbulent kinetic energy transported to the boundary layer is enlarged.

湍流边界层与层流边界层也不同。和层流一样,湍流时紧贴壁面的一层流体也是静止的。由于黏性力的作用,这一层流体对邻近一层流体产生阻滞作用。在管道入口处,管内的湍流与边界层均未充分发展,边界层极薄,为层流边界层。离管口一段距离后,管内湍流获得发展,主流供给边界层的湍动能增强,边界层内流体质点的横向迁移也相当剧烈,因而

Thereby the transverse transportation of the particles in the boundary layer is enhanced correspondingly and the laminar boundary layer converts to be turbulent boundary layer. Even so there still exists an ultra-thin laminar boundary layer that sticking to wall boundary, which is termed the laminar sub-layer. Once the turbulent flow fully develops, the turbulent boundary layer grows to the pipe center.

7.4.2 Classification of Pipe Flow Resistances

The head loss designations of "major" and "minor" do not necessarily reflect the relative importance of each type of loss. It is often necessary to determine the head loss, h_L, that occurs in a pipe flow so that the energy equation can be used in the analysis of pipe flow problems. The overall head loss consists of the head loss due to viscous effects in straight pipes, termed the major loss and denoted $h_{L\text{-major}}$, and the head loss in various pipe components, termed the minor loss and denoted $h_{L\text{-minor}}$.

$$h_L = h_{L\text{-major}} + h_{L\text{-minor}} \qquad (7.20)$$

(1) Major loss

The flow resistance caused by the internal friction between the fluid layers is defined as the on-way resistance, which is also termed the frictional force. This is the so-called major loss in pipe flow. In laminar flow, the drag resistance is produced by the viscous friction. But in turbulent flow, only a little of the drag resistance is caused by viscous friction in the boundary layer, and the majority is caused by the migration and fluctuation of molecular groups.

The major loss corresponds to the energy loss which is eventually used to overcome the friction between the solid surface and the fluid, so it is also termed surface resistance. When the fluid flows in a straight uniform circular tube, the energy loss is mainly caused by the on-way resistance.

In laminar flow, major loss is proportional to fluid velocity which varies smoothly from pipe center to pipe wall. The roughness of the pipe surface influences neither the fluid flow velocity nor the friction loss.

In turbulent flow, losses are proportional to the square of fluid velocity. A layer of chaotic eddies and vortices near the wall surface, termed the viscous sub-layer, forms the transition from wall surface to bulk flow. In the laminar sub-layer, the effects of pipe roughness should be considered. It is useful to characterize that roughness as the ratio of roughness height to pipe diameter, the "relative roughness".

(2) Minor loss

The flow resistance induced by local obstacles is termed the local resistance which is classified as the minor loss. The local obstacles include elbow, expansion and contraction pipes and a wide variety of pipeline fittings, such as valves, Tees, etc. As to the cause of the local resistance, the following sections will be described in detail.

For a pipe system that contains many components and a relatively short pipe, minor loss may be larger than major loss. For any system there are additional so-called minor losses arising from:
• Losses due to inlet and outlet
• Sudden expansion and contraction
• The change of velocity at bends, elbows, tees and other fittings
• Valves, open or partially closed and partial obstruction
• The change of velocity due to gradual expansion and contraction

These components interrupt the smooth flow of the fluid and cause additional losses because of the flow separation and mixing. In a typical system with long pipes, these losses are minor compared with the total head loss in the pipes (the major losses) and therefore are regarded as minor losses. Actually the losses may not be so minor. For example, a partially closed valve can cause a greater pressure drop than that exists in a very long pipe. Minor losses are commonly measured experimentally and correlate with pipe flow parameters. The losses especially for valves are somewhat dependent on the design and manufacturing of product.

While fluid flows in engineering equipment, these two types of flow resistances will be produced simultaneously. So it is necessary to understand the principle of head losses and master the calculation method of fluid resistance.

这些损失相对较小，因此被认为是次要损失。这些损失也不完全是较小的，比如部分关闭的阀门会产生十分大的压降，这个损失要比长管内的沿程损失要大得多。局部阻力损失一般通过实验方法确定，而且与流动参数有关。对于阀门，局部阻力损失与产品的设计和制造有关。

流体在工程设备中流动时，上述两类流动阻力都会产生。因此掌握流体阻力的计算原理和方法是十分必要的。

7.4.3 Calculation of Major Head Loss

The head loss in a long, straight pipe with the diameter of d and length of l will be introduced in this section. For a steady flow of incompressible fluid in smooth pipe, the similarity laws governing the flow include the Reynolds number, Re, the geometrical parameter, l/d, and the Euler number, Eu. For rough pipe, the dimensionless parameter δ/d should also be considered, where δ the roughness of wall surface. According to the similitude theorem, the pressure head loss, Δp, holds the following relationships with these similarity laws.

7.4.3 主要损失的计算

本节讨论流体在直径为 d、长度为 l 的直管内流动时沿程阻力的计算方法。对于光滑管内不可压缩黏性流体的稳定流动，相似准则有雷诺准则 Re、几何准则 l/d 和欧拉准则 Eu；对于粗糙管，还有一个涉及粗糙度的相似准则 δ/d（δ 为绝对粗糙度）。根据相似理论，Δp 与相似准则间有如下关系。

For smooth pipe

对于光滑管有

$$Eu = f\left(Re, \frac{l}{d}\right) \quad (7.21)$$

$$\Delta p = f\left(Re, \frac{l}{d}\right)\rho \bar{v}^2 \quad (7.22)$$

For rough pipe

对于粗糙管，有

$$Eu = f\left(Re, \frac{l}{d}, \frac{\delta}{d}\right) \quad (7.23)$$

$$\Delta p = f\left(Re, \frac{l}{d}, \frac{\delta}{d}\right)\rho \bar{v}^2 \quad (7.24)$$

It has been validated that the head loss, Δp, is proportional to the pipe length, therefore for smooth pipe, Δp can be expressed as

已知，沿程阻力 Δp 与管长 l 成正比。因此对于光滑管，Δp 可表示为

$$\Delta p = 2\varphi(Re)\frac{l}{d}\frac{\rho}{2}\bar{v}^2 \quad (7.25)$$

For rough pipe

对于粗糙管，有

$$\Delta p = 2\varphi\left(Re,\frac{\delta}{d}\right)\frac{l}{d}\frac{\rho}{2}\bar{v}^2 \quad (7.26)$$

Define the terms $2\varphi(Re)$ and $2\varphi(Re, \delta/d)$ as the drag coefficient, λ, then for smooth pipe

将 $2\varphi(Re)$ 和 $2\varphi(Re, \delta/d)$ 定义为阻力系数 λ，那么对于光滑管

$$\lambda = 2\varphi(Re) \quad (7.27)$$

and for rough pipe

对粗糙管，有

$$\lambda = 2\varphi\left(Re,\frac{\delta}{d}\right) \quad (7.28)$$

Then the general equation for Δp can be rewritten as

Δp 的统一计算公式为

$$\Delta p = \lambda \frac{l}{d}\frac{\rho}{2}\bar{v}^2 \quad (7.29)$$

The drag coefficient, λ, should be determined before calculating Δp. Different flow status corresponds to different λ. For smooth pipe, λ is related to Re, but for rough pipe, it is related to Re and relative roughness. So the determination of λ depends upon theoretical and experimental methods jointly, and mainly upon experimental result.

计算 Δp 必须先确定摩擦阻力系数 λ。不同流动状态的 λ 不同。光滑管中 λ 与 Re 有关，粗糙管中 λ 与 Re 和相对粗糙度有关。因此 λ 的确定只能靠理论分析与实验相结合，并且主要依赖于实验结果。

(1) Drag coefficient λ for laminar flow

(1) 层流流动时的阻力系数 λ

The frictional resistance in laminar flow is induced by the viscous friction in fluid. The average velocity in circular pipe is

层流的阻力是由流体的黏性摩擦引起的。圆管中平均流速为

$$\bar{v} = \frac{1}{8\mu}\frac{\Delta p}{l}R^2$$

then

因此有

$$\Delta p = \frac{8\mu l}{R^2}\bar{v} = \frac{32\mu l}{d^2}\bar{v} = \frac{64\mu}{\rho d \bar{v}}\frac{l}{d}\frac{\rho}{2}\bar{v}^2$$

$$\Delta p = \frac{64}{Re}\frac{l}{d}\frac{\rho}{2}\bar{v}^2 \qquad (7.30)$$

| Comparison of Eq. (7.28) and Eq. (7.29) gives | 对比式(7.28)和式(7.29)，有 |

$$\lambda = \frac{64}{Re} \qquad (7.31)$$

(2) Drag coefficient λ for turbulent flow

（2）湍流流动时的阻力系数 λ

According to the turbulent flow velocity equation [Eq. (7.17)],

根据湍流速度方程[式(7.17)]，

$$\frac{v_x}{\sqrt{\tau/\rho}} = 5.75\lg\frac{y\sqrt{\tau/\rho}}{\nu} + 5.5$$

the drag coefficient λ for turbulent flow is in the form

得到湍流时摩擦阻力系数 λ 为

$$\frac{1}{\sqrt{\lambda}} = 2.03\lg\left(Re\sqrt{\lambda}\right) - 1.02 \qquad (7.32)$$

A more exact equation was obtained by experimental investigation, as

根据实验结果，得到了更加精确的计算公式

$$\frac{1}{\sqrt{\lambda}} = 2\lg(Re\sqrt{\lambda}) - 0.8 \qquad (7.33)$$

Equation (7.33) is appropriate when $Re = 3\times10^3 \sim 10^8$. Since it is a transcendental equation, we can use a tentative calculation method to solve λ from the equation.

该式适用于 $Re = 3\times10^3 \sim 10^8$。这是一个超越方程，可用试算法算出 λ 的值。

Furthermore the Blasius empirical formula, as shown by Eq. (7.34), can also be used to calculate λ

此外，还可用勃拉修斯经验公式，即式(7.34)，计算 λ

$$\lambda = \frac{0.3164}{Re^{0.25}} \qquad (7.34)$$

it is appropriate when $Re \leqslant 10^5$ and in this range, Δp is proportional to $\bar{v}^{1.75}$.

该式适用于 $Re \leqslant 10^5$ 的情况，在此范围内，Δp 与 $\bar{v}^{1.75}$ 成正比。

(3) Effects of wall roughness and Re on pressure head loss—the Nikuradse experiment

（3）管壁粗糙度和雷诺数对沿程阻力的影响——尼古拉兹实验

Both the drag coefficients for laminar and turbulent flows are obtained for smooth pipe. But most pipes in

以上阻力系数计算公式基于光滑壁面的假设，而工程中绝对光滑

engineering application are not smooth and the roughness effect on the drag coefficient should be considered.

的管子是不存在的，管壁粗糙度对流动阻力的影响需要考虑。

Figure. 7.10 illustrates the profile of a rough wall surface with the roughness of δ. The symbol Δ is used to represent the relative roughness, i.e., $\Delta = \delta/d$, where d is inner diameter of the pipe.

图 7.10 表示一粗糙度为 δ 的粗糙壁面。用 Δ 表示管壁的相对粗糙度，$\Delta = \delta/d$，其中 d 为管子的内径。

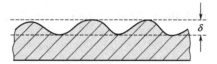

Fig. 7.10 Roughness of pipe surface
图 7.10 管壁的粗糙度

It is not easy to determine the functional dependence of the friction factor on the Reynolds number and relative roughness. Much of this information is a result of experiments conducted by Nikuradse and amplified by many others since then.

要建立雷诺数和相对粗糙度与摩擦系数的方程是很困难的。相关的结论是由尼古拉兹等人经过大量的实验研究得到的。

Nikuradse prepared some pipes with different relative rough surfaces by gluing sand grains of certain size onto pipe walls. The values of the relative roughness, Δ, are 1/30, 1/61.2, 1/120, 1/252, 1/504 and 1/1014, respectively. The pressure drop needed to produce a desired flow rate was measured and the data were converted into the friction factor for the corresponding Reynolds number and relative roughness. The tests were repeated for many times within wide ranges of Re and δ/d to determine the $f = \Phi(Re, \delta/d)$ dependence. The drag coefficient was obtained as shown in Fig. 7.11.

尼古拉兹曾在直径不同的圆管内敷上粒度均匀的沙子，制造出了六种不同相对粗糙度的圆管，Δ 的值分别为 1/30、1/61.2、1/120、1/252、1/504、1/1014。实验测得的压降被转换为与雷诺数、相对粗糙度对应的摩擦系数。经多次重复实验，确定了 $f = \Phi(Re, \delta/d)$ 的关系。阻力系数的实验结果如图 7.11 所示。

When Re is relatively small, $\lg(100\lambda)$ and $\lg Re$ hold the linear relationship, as the line segment AB in Fig. 7.11. The flow is under laminar state and λ only depends on Re. The line segment AB almost coincides with the line of $\lg(100\lambda)$-$\lg Re$ which is obtained from the smooth pipe experiment. It is revealed that, in laminar flow, the flows in rough and smooth pipes hold the same drag coefficient, i.e.

Re 较小时，$\lg(100\lambda)$ 与 $\lg Re$ 呈线性关系，如图 7.11 中线段 AB 所示。流动处于层流状态，λ 仅与 Re 相关。AB 线与光滑管中层流的 $\lg(100\lambda)$-$\lg Re$ 关系曲线几乎重合。也就是说，在层流范围内，粗糙管与光滑管的摩擦阻力系数相同，即

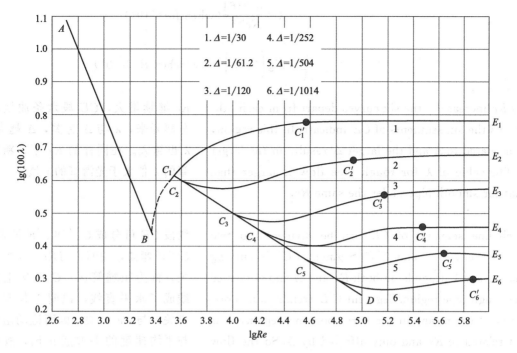

Fig. 7.11 Relationship between Re and λ for rough pipes obtained by Nikuradse
图 7.11 尼古拉兹得到的粗糙管内 Re 和 λ 的关系

$$\lambda = \frac{64}{Re}$$

The critical value of Re for laminar flow in rough pipe ranges from 2160 to 2410, and it is approximately equivalent to 2320 of smooth pipe.

粗糙管的临界雷诺数范围是从 2160 至 2410，与光滑管的数值（2320）大致相等。

The line segment BC corresponds to the transition region from laminar flow to turbulent flow. No obvious regularity can be observed in this range.

BC 段对应着层流转变为湍流的过渡区，该区域中 λ 无明显变化规律。

Corresponding to line segment CD, the flow becomes turbulent and in this region, the six curves trend to be straight and coincide with each other. This result states that λ is only related to Re but not affected by Δ. The slope of line segment CD is -0.25, that is, λ is inversely proportional to the $Re^{0.25}$, which is in accordance to the Blasius formula. In this region, the drag coefficient for rough pipe is the same as that for smooth pipe, and it is defined as the smooth pipe region. λ can be calculated according to

CD 段对应的流动已进入湍流范围，该区域中 6 条线重合，接近一条直线。这说明 λ 与相对粗糙度 Δ 无关，只与 Re 有关；并且 CD 线的斜率为 -0.25，即 λ 与 $Re^{0.25}$ 成反比，符合勃拉修斯公式。此时粗糙管的摩擦阻力系数与光滑管一样，该段称为光滑管区。λ 可根据下式计算

$$\lambda = \frac{0.3164}{Re^{0.25}} \text{ (when } Re < 10^5\text{)}$$

$$\frac{1}{\sqrt{\lambda}} = 2\lg\left(Re\sqrt{\lambda}\right) - 0.8 \text{ (when } Re > 10^5\text{)}$$

As Re increases, the six curves depart from each other, as the six segments of CE indicate. In this region, λ is related to Δ and the larger Δ results in the larger λ. The value of λ for rough pipe flow is larger than that for smooth pipe under the same Re.

Re 继续增大，CE 段六条曲线才分离开来，λ 与 Δ 有关，Δ 越大，λ 也越大。在同样的 Re 下，粗糙管的 λ 值大于光滑管的 λ 值。

Each line segment of CE can be divided into two parts, i.e., CC' and $C'E$. Segment CC' is curving that means λ depends on both Re and Δ and is termed the rough pipe region. Segment $C'E$ trends to be horizontal, that means in the corresponding region, λ is not related to Re and only affected by Δ. So the flow resistance is proportional to the square of the average velocity, and it is defined as the region of quadratic resistance law, i.e., the second self-modeling region that introduced in Chapter 6.

线段 CE 可分成 CC' 和 $C'E$ 两段。CC' 段弯曲，说明 λ 与 Re、Δ 均有关，称为粗糙管区；$C'E$ 段上曲线成了水平直线，说明 λ 仅与 Δ 有关，与 Re 无关，因而流动阻力与平均速度的平方成正比，故称之为阻力平方区，即第 6 章中所说的第二自模化区。

In the region of quadratic resistance law, the equation for λ is established based on experimental results, as shown by

根据实验结果得到的阻力平方区中摩擦阻力系数 λ 的计算式为

$$\frac{1}{\sqrt{\lambda}} = 2\lg\frac{1}{2\Delta} + 1.74 = -2\lg\frac{\delta}{3.7d} \quad (7.35)$$

In the rough pipe region, λ can be calculated according to the Colebrook formula

在粗糙管区中，λ 可由科尔布鲁克（Colebrook）公式计算

$$\frac{1}{\sqrt{\lambda}} = -2\lg\left(\frac{2.51}{Re\sqrt{\lambda}} + \frac{\delta}{3.7d}\right) \quad (7.36)$$

Equation (7.36) is valid for both smooth pipe region and the region of quadratic resistance law. Since λ is not related to Δ for smooth pipe, Eq. (7.36) can be simplified as

上式对光滑管区和阻力平方区都适用。因为在光滑管区，λ 与 Δ 无关，则上式变成

$$\frac{1}{\sqrt{\lambda}} = -2\lg\frac{2.51}{Re\sqrt{\lambda}} = 2\lg(Re\sqrt{\lambda}) - 0.8$$

This is the equation of λ for smooth pipe when $Re > 10^5$. In the region of quadratic resistance law, Re is very large, and $\dfrac{2.51}{Re\sqrt{\lambda}}$ is much smaller than $\dfrac{\delta}{3.7d}$ thus can be ignored. Therefore

$$\frac{1}{\sqrt{\lambda}} = -2\lg \frac{\delta}{3.7d} \qquad (7.37)$$

This is the equation of λ for the region of quadratic resistance law.

Among these equations of λ, which one to be used depends on the resistance area to which a flow corresponds. That is to say the turning points C and C' on line section of CE should be located firstly.

Based on the experimental results, an empirical equation for the critical value of Re at point C' was obtained as

$$Re = 316\left(\frac{1}{2\Delta}\right)^{0.85} \qquad (7.38)$$

where $\dfrac{1}{1014} \leqslant \Delta \leqslant \dfrac{1}{30}$.

The critical value of Re at point C' is

$$Re = 4160\left(\frac{1}{2\Delta}\right)^{0.85} \qquad (7.39)$$

where $\dfrac{1}{1014} \leqslant \Delta \leqslant \dfrac{1}{30}$.

The partition of the resistance region of a flow can be generalized as:

① When $Re \leqslant 2000$, the flow falls in laminar region.

② When $2000 < Re \leqslant 316\left(\dfrac{1}{2\Delta}\right)^{0.85}$, the flow falls in smooth pipe region.

③ When $316\left(\frac{1}{2\Delta}\right)^{0.85} < Re \leqslant 4160\left(\frac{1}{2\Delta}\right)^{0.85}$, the flow falls in rough pipe region.

④ When $Re > 4160\left(\frac{1}{2\Delta}\right)^{0.85}$, the flow falls in the region of quadratic resistance domain.

The roughness of a commercially available pipe is not as uniform and well defined as in the artificially roughened pipes. However, we can measure the effective relative roughness of some typical pipes and thus obtain the friction factor. The effective roughness values of some commercial pipe surfaces are given in Table 7.1.

③ 当 $316\left(\frac{1}{2\Delta}\right)^{0.85} < Re \leqslant 4160\left(\frac{1}{2\Delta}\right)^{0.85}$ 时，属于粗糙管区。

④ 当 $Re > 4160\left(\frac{1}{2\Delta}\right)^{0.85}$ 时，属于阻力平方区。

市面上管道的粗糙度并不如实验室条件下的均匀，但仍然可以测得典型管材的相对粗糙度进而确定摩擦系数。市面上常见管道的有效粗糙度列于表 7.1 中。

Table 7.1 Effective roughness of some commercial pipes
表 7.1 常见商用管道的有效粗糙度

Pipe type 管材类型	δ/mm	Pipe type 管材类型	δ/mm
Seamless steel tubes 无缝钢管	0.04~0.17	Asphalt coated pipe 涂沥青铁管	0.12
Ordinary steel pipe 普通钢管	0.2	Galvanized iron conduit 镀锌铁管	0.15
New steel tube 新钢管	0.12	Copper and aluminum pipe 铜、铝管道	0.0015
Old steel tube 旧钢管	0.5~1.0	Glass and plastic tube 玻璃、塑料管	0.001
Ordinary cast iron pipe 普通铸铁管	0.5	Rubber hose 橡胶软管	0.01~0.03
New cast iron pipe 新铸铁管	0.25~0.42	Concrete pipe 混凝土管	0.33
Old cast iron pipe 旧铸铁管	1.0~3.0	Ligneous pipe 木材管	0.25~1.25

In nature, the effect of roughness and Re on major head loss lies on the flow field near the rough surface, as can be seen in Fig. 7.12.

实质上，粗糙度和雷诺数对沿程阻力的影响取决于管壁附近流体的流动状况，如图 7.12 所示。

(1) In the laminar flow region

(1) 层流区

The flow is laminar when Re is relatively small. Wall roughness has weak effect on the drag force, because

Re 较小时流动为层流，粗糙度对阻力的影响不明显，流动不会在

the fluid will not separate from the bulges or form eddies on the back of the bulges. Bulges have a little disturbance on the flow and the streamlines easily trend to be straight when they are far away from the rough boundary, as illustrated by Fig. 7.12(a). The pressure head loss is induced by viscous friction among fluid layers and not related to roughness.

凸出物处分离或在其背面形成旋涡。凸出物对流动干扰很小，流线离开粗糙壁面后很快趋于平直，如图 7.12(a) 所示。沿程阻力损失主要由流体层间的摩擦力引起，与粗糙度无关。

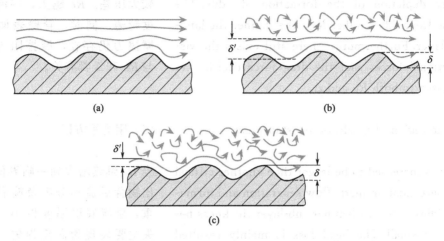

Fig. 7.12　Flow fields near rough surface corresponding to different flow status
图 7.12　粗糙管壁附近流体的流动状况

(2) In the smooth pipe region

(2) 光滑管区

As Re increases but is not too large, the bulk of the flow becomes turbulent. If the bulges are covered by the laminar sub-layer, as Fig. 7.12(b) illustrates, the laminar sub-layer thickness δ' is larger than δ and the roughness has just a little effect on the drag force which is only related to Re, thus it is similar to that in smooth pipe. Since δ' will reduce with the increasing Re, as Re increases further to be a critical value, δ' will be smaller than δ finally that means the bulges will penetrate the laminar sub-layer. Obviously, the larger roughness corresponds to the smaller critical value of Re which corresponds to the flow status that bulges penetrate the laminar sub-layer. That results in the different start points of the line segments of CE shown in Fig. 7.11.

Re 不太大时，主流处于湍流状态，但如果层流底层能淹没管壁的粗糙凸出物，即图 7.12(b) 所示，层流底层厚度 δ' 大于 δ，则粗糙度对阻力的影响很小，阻力仅与 Re 有关，此时类似于光滑管。由于 δ' 随 Re 的增大而减小，所以当 Re 达到一定值时，δ' 会小于 δ，即凸出物会穿破层流底层。显然，粗糙度越大，凸出物伸出层流底层时对应的 Re 越小。这就是图 7.11 中各个 CE 线具有不同起点的原因。

(3) In the rough pipe region

When Re is relatively large, $\delta > \delta'$, the flow around bulges is turbulent and flow separation occurs at the back of bulges accompanied by the formation of eddies [Fig. 7.12(c)]. Large pressure drop is generated between the windward and leeward sides of a bulge due to energy depletion of the formation of eddy. The larger the Reynolds number Re, the thinner the laminar sub-layer. Furthermore, there still exists the viscous frictional force, and the drag force coefficient, λ, is related to both Re and Δ.

(4) In the region of quadratic resistance law

When Re is increased to be larger than another critical value, there exist dramatic flow separation and formation of eddies, and the laminar sub-layer thickness becomes very small. The head loss is mainly resulted from eddy losses. Comparatively the head loss due to viscous frictional force is much smaller than turbulent loss. Therefore, it can be regarded that λ will not vary with Re and the flow falls in the region of quadratic resistance law.

The difficulty with Eq. (7.36) is that it is implicit in the dependence of λ, therefore a sort of iterative scheme should be conducted to solve λ from the equation. Figure 7.13, the Moody chart, shows the functional dependence of λ on Re and Δ, which is obtained by L. F. Moody from the Colebrook equation [Eq. (7.36)]. In fact, the Moody chart is a graphical representation of Eq. (7.36), which is an empirical fit of the pipe flow pressure drop data.

7.4.4 Calculation of Minor Head Loss

Because the fluid is needed to be turned, regulated, measured, filtrated, accelerated, pressurized and the like, most pipe systems used in engineering consist of

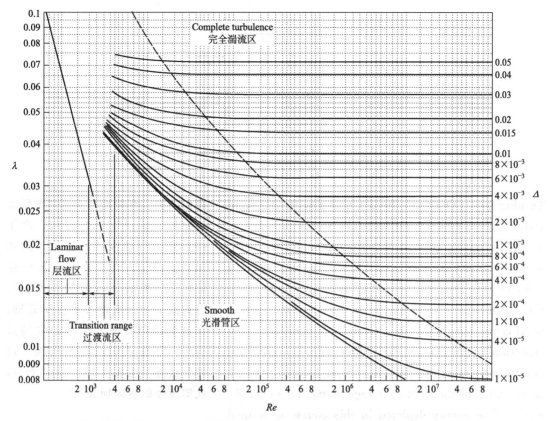

Fig. 7.13 Drag force coefficient λ as a function of Re and Δ—the Moody chart

图 7.13 阻力系数 λ 与 Re 和 Δ 的关系——莫迪图

additional components, such as bend, tee, valve, flowmeter, filter, reducer, etc., except straight pipe. These components add minor losses to the overall head loss of a pipe flow. In this section the determination of various minor losses will be indicated.

还包括弯头、三通、阀门、流量计、过滤器和变径段等管路附件，管路附件将增加流动的阻力损失。本节介绍局部阻力的确定方法。

(1) Classification of minor losses

（1）次要损失的分类

The minor losses can be classified as collision loss, turning loss, eddy loss and transmission loss. These minor losses can be seen in the flow through a sudden shrinking passage, as Fig. 7.14 illustrates.

次要损失有四种类型：碰撞损失、转向损失、涡流损失和变速损失。图 7.14 所示的突然收缩管中出现了这 4 种损失。

Collision loss As a fluid flows from cross section 1 to section 2, a part will collide with the solid wall at cross section 2. Considering the real fluid is not an ideal elastic body, there must be the energy loss during the collision process.

碰撞损失 流体由 1 截面向前流动，一部分流体在截面 2 处与管壁发生碰撞。实际流体不是理想弹性体，碰撞的结果必定会产生能量损失。

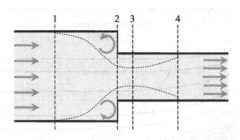

Fig. 7.14 Minor losses occurring in sudden shrinking passage
图 7.14 突然收缩管道内的次要损失

Turning loss Once the fluid collides with the solid wall at cross section 2, it will turn to the pipe center and there exist a vertical velocity component with respect to the pipe centerline (i.e., radial velocity component). Due to the radial velocity component, the neckdown of streamlines occurs at cross section 3. At cross section 4, the streamlines parallel to the centerline again. During the course of the occurrence and disappearance of the radial velocity component, there exists momentum exchange along the radial direction. The energy depleted in this course is termed the turning loss.

转向损失 在截面 2 上流体与壁面碰撞后，转向管中心方向，产生一垂直于管轴线的分速度，即径向速度分量。由于径向速度分量的作用，截面 3 处流线"颈缩"，直到截面 4 处流线又平行于管轴。在这个径向流速产生和消失的过程中，存在径向的动量交换。这种交换显然会损失掉一部分能量，此即流体的转向损失。

Eddy loss Accompanied by the shape change of streamlines, the separation of bulk flow from wall boundary, as well as the formation of eddies, always occur at cross sections 2 and 3. The eddy loss arises from two aspects, one is the mechanical energy loss induced by viscous frictions both in eddies and between the eddy and wall boundary; the other is the energy loss due to the momentum exchange between eddies and bulk flow.

涡流损失 在流线形状发生突变处，往往会造成主流脱离边界壁面并形成旋涡，如图 7.14 中截面 2、截面 3 处。涡流损失来自于两个方面，一是涡流内部及涡流与壁面间的黏性摩擦消耗的机械能；二是旋涡与主流的动量交换造成的能量损失。

Transmission loss From cross sections 1 to 3, the flow velocity increases while the pressure decreases, and a partial pressure energy converts to kinematic energy. But from cross sections 3 to 4, the velocity reduces and the pressure increases reversely, and partial kinematic energy converts to pressure energy. As we know, the efficiency of energy conversion is usually smaller than 100%, especially during the course of

变速损失 流体从截面 1 到截面 3 的流动是加速降压过程，部分压力能转变为动能；从截面 3 到截面 4 的流动是减速扩压过程，部分动能转变为压力能。这种能量转变显然不会有 100% 的效率，特别是减速扩压段，即由截面 3 到截面 4，能量损失较为显著，严重时会引

pressure increasing and velocity decreasing (from cross section 3 to section 4) that may induce both the separation of bulk flow from wall boundary and inverse flow. This is the generation of transmission loss.

These four minor losses only occur in local regions of the flow passage, so they are also termed the local losses. Of course there also exists head loss due to viscous friction in these local regions, which is much smaller than the minor losses and usually ignored.

(2) Head losses due to flow passage change

Similar to that in sudden shrinking pipe, the flows in sudden expansion pipe and bending pipe, as well as the flow around an object, also encounter minor head losses.

In sudden expansion pipe As illustrated in Fig. 7.15, eddies form in the corner region behind cross section 1. So the eddy loss definitely exists. Furthermore, as the fluid flows into a larger space, the velocity is redistributed, which also results in the transmission loss.

起主流脱离壁面，甚至产生回流。这种由速度变化引起的损失称为变速损失。

这四种次要损失只发生在局部区域，又称为局部损失。当然这些局部结构处同样存在黏性阻力损失，但比局部阻力小得多，可忽略不计。

（2）流道变化产生的损失

与管道截面突然缩小一样，管道截面突然扩大、管道弯曲、绕物流动时均会产生局部损失。

管道截面突然扩大 如图 7.15 所示，在截面 1 后的角形区域形成了涡流，造成涡流损失；同时流体流入大截面后速度重新分布，带来变速损失。

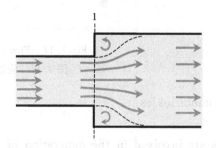

Fig. 7.15 Sudden expansion of flow passage
图 7.15 突然扩大管道

In bending pipe Due to inertia force, the pressure increasing and decreasing regions coexist in an elbow flow as shown in Fig. 7.16. Streamlines separates while passing through the elbow and result in the formation of eddies, thus the eddy loss occurs. Furthermore, the pressure difference between the two regions generates a backflow in the elbow. The confluence of the backflow and the bulk flow forms a spiral flow, and that also results in a large local head loss.

管道弯曲 如图 7.16 所示，流体流过弯管时，由于惯性力作用，在弯道内外侧同时形成增压和减压区。此处往往产生流线分离形成旋涡，造成涡流损失；同时，由于内外侧的压差，弯管处产生回流，附加在主流上形成了螺旋流动，从而造成较大的局部损失。

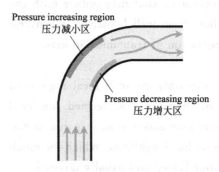

Fig. 7.16　Flow through a bending pipe
图 7.16　弯管内的流动

Flow around obstacles　When the fluid encounters valve, gate or other objects in a pipe system, it will collide with these obstacles, causing the collision loss. On the other hand, the boundary layer separates from the obstacle surface and vortex region forms at the leeward surface of obstacle, as shown in Fig. 7.17, resulting in the eddy loss.

流体绕过物体的流动　如图 7.17 所示，流体在管道内流动遇到阀门、闸板等物体时，与这些物体发生碰撞，引起碰撞损失。另一方面物体表面的边界层会产生脱离现象，结果在物体后部产生涡流区，造成涡流损失。

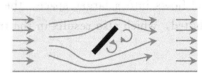

Fig. 7.17　Flows around obstacles
图 7.17　绕过物体的流动

(3) Calculation of minor losses

(3) 次要损失的计算

Since many factors are involved in the generation of minor losses which occur in the irregular regions of flow passage, minor losses are usually determined by experimental method. The minor pressure head loss can be expressed as

影响流道不规则区域中的次要损失的因素很多，因此，次要损失的大小一般通过实验来确定。次要损失可用下式表示

$$\Delta p = \xi \frac{\rho v^2}{2} \quad (7.40)$$

where the drag coefficient, ξ, is experimentally determined and can be consulted in handbook. The drag coefficients of some typical minor losses are introduced below.

式中，阻力系数 ξ 一般由实验测定，可以从专门的手册中查到。下面简单介绍几种典型的局部阻力系数。

In sudden shrinking pipe The drag coefficient ξ is

$$\xi = 0.5\left(1 - \frac{A_2}{A_1}\right)$$

where A_1 and A_2 are the areas of large and small cross sections, respectively. When a liquid flows from a big container into a small pipe, $A_1 \to \infty$ and $A_2/A_1 \approx 0$, $\xi = 0.5$.

In sudden expansion pipe The drag coefficient ξ is

$$\xi = \left(1 - \frac{A_1}{A_2}\right)^2$$

where A_1 and A_2 are the small and large cross section areas, respectively. When the fluid flows from a small pipe into a big container (Fig. 7.18), i.e., $A_2 \to \infty$ and $A_1/A_2 \approx 0$, $\xi = 1$. On this condition, the velocity of the fluid in the big container becomes zero, and all the dynamical energy is depleted.

管道截面突然收缩 ξ 为

式中，A_1 和 A_2 分别为大、小端的面积。当液体由大容器流入小管道，$A_1 \to \infty$，$A_2/A_1 \approx 0$，$\xi = 0.5$。

管道截面突然扩大 ξ 为

式中，A_1 和 A_2 分别为小、大端的面积。当流体由小管道流入大容器，如图 7.18 所示，$A_2 \to \infty$，$A_1/A_2 \approx 0$，$\xi = 1$。此时可认为大容器中的流速为 0，说明流体损失了全部动能。

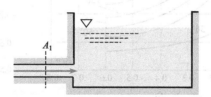

Fig. 7.18 Flow from small pipe into big container
图 7.18 流体由小管道流入大容器

In bending pipes Bending pipes include gentle bend and sharp bend, and their sketch maps can be seen in Fig. 7.19(a) and (b), respectively, and Fig. 7.19(c) shows a bend with discontinuous transition. The experimental results of the drag coefficients for some bends are also shown in Fig. 7.19. For alloy-steel pipes with the diameter larger than 150mm, as well as any austenitic steel pipes, the drag coefficient should be multiplied by a correction factor, C_Δ, which is given in Fig. 7.19(d) where the curves 1, 2 and 3 correspond to the three pipes shown in Fig. 7.19(a), (b) and (c), respectively.

弯管 管道上的弯头分缓转弯头和急转弯头两种，如图 7.19(a) 和 (b) 所示，图 7.19(c) 所示为非连续过渡弯管。实验测定的一些弯头的阻力系数如图 7.19 所示。若是直径大于 150mm 的合金钢管或任意直径的奥氏体钢管，阻力系数应乘以一修正系数 C_Δ，该系数由图 7.19(d) 给出，其中曲线 1、2、3 分别对应图 7.19(a)、(b)、(c) 所示三种型式的弯管。

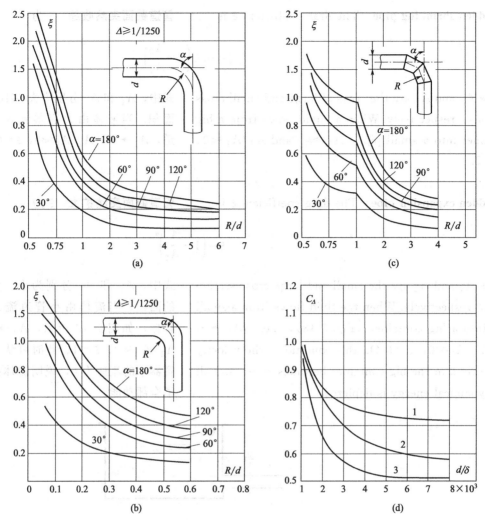

Fig. 7.19 Drag coefficients for some bending pipes

图 7.19 一些弯管的阻力系数

In valves The drag coefficient of a valve is usually large and depends on its opening degree. In most valves, collision loss, turning loss, eddy loss and transmission loss exist jointly. These drag coefficients can be queried in certain handbooks.

(4) Methods to reduce minor losses

Minor losses will increase the energy consumption of pipe system. Therefore it is necessary to minimize the minor losses in the design of fluid machinery and pipe systems. Some frequently-used methods will be introduced below.

阀门 阀门的阻力系数一般较大，且与阀门开度有关。大多数阀门中，碰撞、转向、涡流和变速损失都存在，具体数据可查手册。

(4) 减小次要损失的措施

次要损失会增加管路系统的能量损失，因此流体机械和管道设计时应充分考虑如何减小局部阻力。常见方法介绍如下。

The entry of a pipe is suggested to be rounding off For the sudden shrinking pipe as shown in Fig. 7.14, the maximum drag coefficient ξ is 0.5 for a flow from large pipe to small pipe. But if the inlet of small pipe is rounded off, the drag coefficient can be reduced remarkably. For common smooth entry, $\xi = 0.1 \sim 0.2$, and for streamline entry (Fig. 7.20), ξ is so small that it can be neglected.

管道入口采用圆滑过渡 图 7.14 所示的突然收缩管道中，最大阻力系数 $\xi=0.5$。如果把小管入口处圆滑过渡，则阻力系数可显著减小。对于一般匀滑入口，$\xi=0.1\sim0.2$；对于流线型入口（图 7.20），ξ 可小至忽略不计。

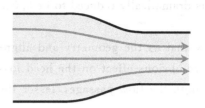

Fig. 7.20 Streamlined entry
图 7.20 流线型入口

Head loss at pipe inlet completely depends on the geometry and alignment of the pipe because different alignment results in different head loss coefficient. The connection between a tank and a pipe may have many schemes, as shown in Fig. 7.21, the geometry and connection position of the pipe affects the head loss coefficient ξ.

管路入口损失与入口处的几何形状及入口处的连接方式有关。接管与水槽的连接有很多种形式，接管的几何形状及接入位置将会影响入口阻力系数 ξ，如图 7.21 所示。

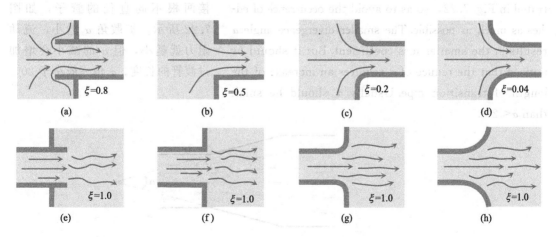

Fig. 7.21 Drag coefficients of some pipe outlets and inlets
图 7.21 一些流入管道和流出管道结构的阻力系数

In case 1 [Fig. 7.21(a)], the pipe connected to the tank is slightly inside the tank and this alignment disturbs the flow around the entrance thus the loss coefficient is as

第 1 种情况 [图 7.21(a)]，接管略微伸入水槽中，这使得接管处的流动不流畅，阻力系数高达 0.8。

high as 0.8. In case 2, [Fig. 7.21(b)] the pipe is attached to the tank through the wall but the joint is acute. So in this case the head loss coefficient is 0.5. The alignment in case 3 [Fig. 7.21(c)] is same as that in case 2 but the joint are rounded which smooth the flow, and ξ is reduced to be 0.2. In case 4 [Fig. 7.21(d)] the pipe alignment is also the same as that in case 2 and case 3 but the joint of the pipe are more rounding. As a result ξ is dramatically reduced to be 0.04.

The four cases show that as the geometry and alignment of the pipe have obvious effect on the head loss coefficient. More smooth the flow passage, lesser the head loss coefficient.

But for the cases of fluid flows from a pipe into a tank, as cases 5 to 8 [Fig. 7.21(e)~(f)], the head loss coefficient is enlarged to be 1 regardless of geometry or alignment of the pipe.

Connection of pipes with different diameters using gradient pipe The connection of two pipes with different diameters is suggested to use a gradient pipe, as illustrated in Fig. 7.22, so as to avoid the occurrence of eddies as much as possible. The smaller divergence angle α results in the smaller loss coefficient. But it should be noticed that the reduce of α requires an increase of the length of transition pipe. Usually α should be small than $\alpha < 20°$.

第 2 种情况［图 7.21(b)］,接管口与槽壁连接,但入口处是尖锐的,此时阻力系数为 0.5。第 3 种情况［图 7.21(c)］,将第 2 种接管入口处进行了圆滑过渡,ξ 减小到 0.2。第 4 种情况［图 7.21(d)］中管子的承插方式与 2 和 3 类似,但过渡更加圆滑,阻力系数减小为 0.04。

由以上实例可知,接管的几何结构和承插方式对阻力系数影响显著。管路越流畅,阻力系数越小。

对于第 5 至第 8 种情况［图 7.21(e)~(h)］,流体从管道流入水槽,阻力系数均增大为 1 且与接管的几何形式和位置无关。

采用锥形管连接异径管 为尽量避免截面突变产生旋涡,常用截面逐渐变大或变小的锥形管来连接两根不同直径的管子,如图 7.22 所示。扩散角 α 越小,流动阻力就越小,但 α 的减小会增加过渡管的长度。α 角一般小于 20°。

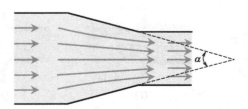

Fig. 7.22 Gradient pipe between two pipes with different diameters
图 7.22 采用渐变管连接异径管

Optimization of flow in bends The local resistance of the flow in a bending pipe is related to the pipe diameter, bending radius and central angle β (shown in

优化弯头中的流型 弯管的局部阻力与中心角、管子直径及弯曲半径有关。对小直径管道,应尽可

Fig. 7.23). For a pipe with small diameter, β should be increased as much as possible. In a pipe with large diameter, eddies are more easily to occur at the turning of the elbow, and the secondary flow will also occur at this position. By fixing the guide vanes made of thin steel plate in the elbow (Fig. 7.24) as well as increasing the central angle, the local head loss coefficient ξ can be reduced from 1.1 to 0.4. If meniscate guide vanes are used, ξ can be further minimized to be 0.25.

能增大中心角 β，如图 7.23 所示。对大直径管道，由于弯管内容易产生涡流，截面上又易产生二次流，因此在增大弯曲半径的同时，可在图 7.24 所示的弯道内安装导流板，可将阻力系数 ξ 从 1.1 降至 0.4。若采用月牙形导流板，阻力系数可进一步降至 0.25。

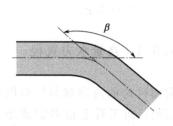

Fig. 7.23　A bending pipe section
图 7.23　弯管段

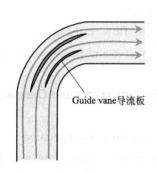

Fig. 7.24　Flow in bending pipe with meniscate guide vanes
图 7.24　带月牙形导流板的弯管中的流动

Optimization of flows in Tees　Tees are frequently used in pipe system to converge flows or split a flow. In this situation, guide vanes can also be used to reduce flow resistance. For example, when two flows converge to a pipe which is perpendicular to the incoming flows (Fig. 7.25), the guide vane can make the two flows turn steadily.

三通内流型的优化　三通常用于并流或分流。为减少流体通过三通时的阻力，可在总管中安装导流板。如图 7.25 所示，当两股流体向垂直的总管汇合时，导流板可使流体转向更加平稳。

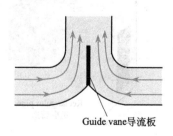

Fig. 7.25　Convergence of two flows in a Tee with guide vane
图 7.25　带导流板的三通中的流体汇合

7.5 Calculation of Head Loss in Pipeline

Calculation of head loss in pipeline is of great importance in application of fluid mechanics, the purpose of which is to determine the relationships among flow rate, pipe size and flow resistance. The head loss calculation include:

① Determine the pipe diameter according to the given flow rate and permissible head loss.
② Check the head loss according to the given pipe diameter and flow rate.
③ Calculate the flow rate for the given pipe with known pipe diameter and permissible head loss.

7.5.1 Equivalent Hydraulic Diameter

The former discussions are specified for circular pipes. In engineering, many conduits are not circular in cross section. As we know, the velocity distribution is axisymmetric in circular pipe. But for an arbitrary cross section, as shown in Fig. 7.26, the velocity profile is a function of both coordinates x and y. Although the details of the flows in such conduits depend on the exact cross-sectional shape, many flow resistance conclusions for circular pipes can be consulted and applied, with slight modification, to flow in conduits with other shapes.

7.5 管路损失计算

管路压头损失计算是流体力学的一个重要应用,可确定流速、管路尺寸和流动阻力之间的关系。压头损失计算包括:

① 根据给定的流量和允许的压降确定管道直径。
② 根据给定的管道直径和流量来核算压降。
③ 根据给定的管道直径、允许的压降,校核流量。

7.5.1 当量水力直径

前面讨论的内容都是针对圆形管道的,而工程上很多管道是非圆截面的。圆管内的流速是呈轴对称分布的,而非圆形管道(图7.26)速度的分布与流动截面的两个坐标都有关。虽然非圆形管道中的流动特征与截面形状有关,但是圆管流动阻力的结论可用于其他管道,不过需要对某些参数进行修正。

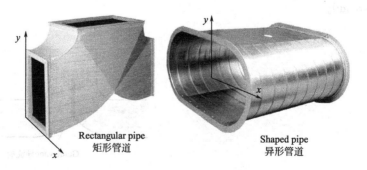

Fig. 7.26 Conduits with noncircular cross sections
图 7.26 非圆截面管道

From the point of similitude, the drag coefficient, λ, 根据相似理论,圆管的摩擦阻力系

proposed for circular pipe flow is not appropriate for noncircular pipe flow because they are not similar in geometry. But experimental research revealed that the dominant factor in drag coefficient is Re while the pipe shape is the secondary influencing factor. Therefore for turbulent flow, if Re is not too large, Eq. (7.40) can also be used to calculate the drag coefficient for noncircular pipes with the diameter replaced by the equivalent hydraulic diameter, d_e, while calculating the Reynolds number. d_e is defined as

$$d_e = \frac{4A}{L} \quad (7.41)$$

where A is the flow area and L is the length of the wetted periphery.

7.5.2 Head Loss Calculation of Pipe System

(1) Head loss in single pipe

The flow in a single pipe (with constant cross section and without bypass) follows the mass conservation principle as

$$\dot{m} = \rho Q = \rho v A \quad (7.42)$$

The total pressure head loss, Δp_Σ, is the summation of major and minor losses, which holds

$$\Delta p_\Sigma = \left(\lambda \frac{l}{d_e} + \Sigma \xi\right) \frac{\rho v^2}{2} \quad (7.43)$$

For a specified pipeline, the minor drag coefficient ξ is known, and the major drag coefficient λ is related to both Re and Δ. The fluid viscosity, μ, relative roughness, Δ, and pipe length, l, are also known. The characteristic size of the pipe, d_e, can be calculated according to Eq. (7.41). Thereby the variables in Eq. (7.42) include v, d_e and Δp_Σ, and each one can be determined when the other two are given.

In a type ⅰ problem, v and d_e are specified, we need

数 λ 不适用于非圆截面管道,因为两者几何不相似。但实验结果表明,对 λ 起决定作用的是 Re,截面形状是次要影响因素。对于湍流,若 Re 不太大,式(7.40)可以用于非圆截面管道中的阻力系数,其中 Re 的定性尺寸要按当量直径 d_e 来计算。d_e 定义为

式中,A 是流通截面面积;L 为湿润周边长度。

7.5.2 管道系统的损失计算

(1) 简单管路计算

简单管路是指管道截面不变、无支路,其连续性方程为

管路的总流动阻力等于沿程阻力和局部阻力之和,公式为

对某一管路,次要阻力系数 ξ 已知,主要阻力系数 λ 与 Re 和 Δ 有关。μ 和 Δ 可认为是已知的,管长 l 也是给定的。当量直径 d_e 由公式(7.41)计算。因此,式(7.42)中的变量是 v、d_e 及 Δp_Σ。若已知其中两个,就可求出另外的一个。

对第 ⅰ 种类型的问题,已知 v 和

to find the necessary pressure drop or head loss. Re can be calculated using v and d_e, and λ can be determined according to Eq. (7.36) or the Moody chart. Then Δp_Σ can be calculated using Eq. (7.43).

In a type ⅱ problem, Δp_Σ and d_e are specified and the flow rate in the pipe is to be determined. Firstly suppose an initial value for λ and calculate v and Re. Then find the corresponding λ and check whether it is coincident with the previous value. If it is not, revise the value of λ and repeat the process. Once λ is determined, the average velocity, v, and flow rate, Q, can be obtained.

In a type ⅲ problem, Δp_Σ and Q are specified and the diameter of the pipe is to be determined. The average velocity v can be calculated by $v = Q/A$ (for circular pipe, $v = 4Q/\pi d^2$). Then Eq. (7.42) can be rewritten as

$$\Delta p_\Sigma = \left(\lambda \frac{l}{d} + \Sigma \xi\right)\frac{\rho}{2}\left(\frac{4Q}{\pi d^2}\right)^2 \qquad (7.44)$$

Assume a corresponding flow resistance region for the problem of interest (the flows in engineering usually fall in the region of quadratic resistance law), and the drag coefficient λ is the function of d, i.e., $\lambda = f(d)$. Then Δp_Σ can be expressed by a function of d

$$\Delta p_\Sigma = \phi(d) \qquad (7.45)$$

We can solve d from Eq. (7.44) and then check whether the assumed resistance region is right or not. If it is not, assume another resistance region for the flow.

【EXAMPLE 7.1】 A seamless steel tube with 5000m-length and 250mm-diameter is used to transport fossil oil with the mass flow rate of 100t/h. The needed physical parameters of the oil include $\nu_{winter} = 1.09 \times 10^{-4} \text{m}^2/\text{s}$, $\nu_{summer} = 0.36 \times 10^{-4} \text{m}^2/\text{s}$ and $\rho = 885 \text{kg/m}^3$. Calculate the head losses of the pipe in winter and summer respectively.

d_e，求压降。Re 可根据 v 和 d_e 计算。λ 可根据式(7.36)或莫迪图确定。最终可用式(7.43)直接计算 Δp_Σ。

对第ⅱ种类型的问题，Δp_Σ 和 d_e 已知，可根据经验先假定一个 λ 值，算出 v 和 Re，然后对假定的 λ 值进行校核，如果误差较大，则重新假设 λ，直至 λ 的偏差满足要求为止。最后用准确的 λ 值求得 v 和 Q。

对第ⅲ种类型的问题，Δp_Σ 和 Q 已知，需确定管径。v 可由 $v = Q/A$ 计算得到，对于圆管 $v = 4Q/\pi d^2$，代入式(7.42)得

假定一个阻力区域（在实际中所遇到的流动大多属阻力平方区），则沿程阻力系数 λ 可以写成 $\lambda = f(d)$ 的形式。于是 Δp_Σ 可用一个关于 d 的方程表示

解得 d，然后检验先前假定的阻力区域是否正确。若不正确则重新进行假设和检验。

【例题 7.1】 输送石油的管道为长5000m、直径为 250mm 的无缝钢管，通过管道的质量流量为100t/h。石油物性参数为 $\nu_{winter} = 1.09 \times 10^{-4} \text{m}^2/\text{s}$、$\nu_{summer} = 0.36 \times 10^{-4} \text{m}^2/\text{s}$，$\rho = 885 \text{kg/m}^3$，计算冬夏两季沿程阻力损失。

SOLUTION　The pressure head loss equation is

$$h_f = \Delta p_\Sigma / \rho g = \lambda \frac{l}{d} \frac{v^2}{2g}$$

The volume flow rate is

$$Q = \frac{\dot{m}}{\rho} = \frac{100 \times 10^3}{885} = 113 \, (\text{m}^3/\text{h})$$

The average velocity is

$$v = \frac{4Q}{\pi d^2} = \frac{4 \times 113}{3600 \times 3.14 \times 0.25^2} = 0.64 \, (\text{m/s})$$

Then the Reynolds numbers are

$$Re_{\text{winter}} = \frac{vd}{\nu_{\text{winter}}} = \frac{0.64 \times 0.25}{1.09 \times 10^{-4}} = 1467.9$$

$$Re_{\text{summer}} = \frac{vd}{\nu_{\text{summer}}} = \frac{0.64 \times 0.25}{0.36 \times 10^{-4}} = 4444.4$$

The roughness of seamless steel tube δ is 0.19mm. So the critical Reynolds number is

$$316 \left(\frac{1}{2\Delta}\right)^{0.85} = 78545.3 \quad \text{and} \quad 4160 \left(\frac{1}{2\Delta}\right)^{0.85} = 1034013.6$$

It can be seen that

$$Re_{\text{winter}} < 2000 \quad \text{and} \quad 2000 < Re_{\text{summer}} < 78545.3$$

In winter the flow is laminar and in summer the flow falls in the smooth pipe region. Therefore, in winter

$$\lambda = \frac{64}{Re} = \frac{64}{1467.9} = 0.0436$$

$$h_f = \lambda \frac{l}{d} \frac{v^2}{2g} = 0.0436 \times \frac{5000}{0.25} \times \frac{0.64^2}{2 \times 9.81} = 18.2 \, (\text{meters of oil column})$$

and in summer

$$\lambda = \frac{0.3164}{Re^{0.25}} = \frac{0.3164}{4444.4^{0.25}} = 0.0388$$

$$h_f = \lambda \frac{l}{d} \frac{v^2}{2g} = 0.0388 \times \frac{5000}{0.25} \times \frac{0.64^2}{2 \times 9.81} = 16.2 \, (\text{meters of oil column})$$

【EXAMPLE 7.2】 Find the pressure drop ($p_1 - p_2$) in pascal across the entire pipeline illustrated in Fig. 7.27. The flow rate of water is $0.05 \text{m}^3/\text{s}$ and $\rho = 1 \times 10^3 \text{kg/m}^3$, the pipe diameter is 200mm and the gate valve is half-open. The elevation difference (along the z direction) between the two points is 10m. $\lambda = 0.008$, $\xi_{\text{gate valve}} = 4.5$ and $\xi_{90° \text{ elbow}} = 0.75$.

【例题 7.2】 计算图 7.27 所示的管路系统的压降 $p_1 - p_2$。已知水的流量为 $0.05 \text{m}^3/\text{s}$、密度 $\rho = 1 \times 10^3 \text{kg/m}^3$，管径 200mm，闸阀半开。两个位置的高度差（$z$ 方向）为 10m。$\lambda = 0.008$，$\xi_{\text{gate valve}} = 4.5$，$\xi_{90° \text{ elbow}} = 0.75$。

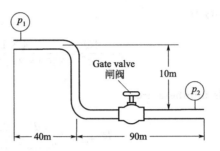

Fig. 7.27   Figure for example 7.2
图 7.27   例题 7.2 图

SOLUTION   The Bernoulli equation for this problem is

解   该问题的伯努利方程为

$$\frac{v_1^2}{2g} + z_1 + \frac{p_1}{\rho} = \frac{v_2^2}{2g} + z_2 + \frac{p_2}{\rho} + h_f$$

where   其中

$$h_f = \left(\lambda \frac{l}{d_e} + \Sigma \xi\right) \frac{v^2}{2g}$$

Since the pipe is of constant diameter, we have $v_1 = v_2$, then

管径不变，因此 $v_1 = v_2$，于是

$$p_1 - p_2 = \rho g (z_2 - z_1) + \left(\lambda \frac{l}{d_e} + \Sigma \xi\right) \frac{\rho v^2}{2g}$$

where $z_2 - z_1 = -10\text{m}$, $\lambda = 0.008$, $l = 140\text{m}$, $d_e = 0.2\text{m}$, $\Sigma \xi = \xi_{\text{gate valve}} + 2 \times \xi_{90° \text{ elbow}} = 6$, and

式中，$z_2 - z_1 = -10\text{m}$；$\lambda = 0.008$；$l = 140$；$d_e = 0.2\text{m}$；$\Sigma \xi = \xi_{\text{gate valve}} + 2 \times \xi_{90° \text{ elbow}} = 6$；且

$$v = \frac{4 \times 0.05}{\pi \times 0.2^2} = 1.6 \text{ (m/s)}$$

The pressure drop can be obtained as   压降为

$$p_1 - p_2 = 1000 \times 9.8 \times (-10) + \left(0.008 \times \frac{140}{0.2} + 6\right) \frac{1000 \times 1.6^2}{2 \times 9.8} = -96485 \text{ (Pa)}$$

The elevation difference between inlet and outlet provide enough hydrostatic pressure to drive the flow. Compared with the potential energy, the head loss in the pipe is much smaller.

【EXAMPLE 7.3】 Water flows in a long pipe with the length of 2438m and flow rate of 1.06m³/s. The pipe is new cast iron pipe and the allowable pressure drop is 63.53mH$_2$O. Determine the diameter of the pipe. $\nu_{water} = 1.0 \times 10^{-6}$ m²/s.

SOLUTION The total pressure head loss is

$$\Delta p_\Sigma = \left(\lambda \frac{l}{d_e} + \Sigma \xi\right) \frac{\rho v^2}{2} = \rho g h_w$$

where h_w presents the height of water column. No minor losses exist in the problem, so $\Sigma \xi = 0$. Then

$$h_w = \lambda \frac{l}{d} \frac{v^2}{2g}$$

i.e.

$$\lambda \frac{2438}{d} \frac{v^2}{2g} = 63.53 \text{ (m)}$$

Substituting $v = Q/A = 1.06/(\pi d^2/4)$ into the above equation gives

$$d^5 = 3.56\lambda$$

Re can be obtained as

$$Re = \frac{vd}{\nu} = \frac{\left(1.06 / \frac{\pi}{4} d^2\right) d}{1.0 \times 10^{-6}} = \frac{1.35 \times 10^6}{d}$$

The roughness of new cast iron pipe ranges from 0.25 to 0.42mm, as can be seen in Table 7.1. Here δ can be valued as 0.3mm. Because d and λ are unknown, we should specify an initial value for λ. Assume $\lambda = 0.02$, we have

$$d = \sqrt[5]{3.56 \times 0.02} = 0.59 \text{ (m)}$$

The relative roughness is $\Delta = \delta/d = 0.3\times 10^{-3}/0.59 = 5.1\times 10^{-4}$. The average velocity is

相对粗糙度为 $\Delta = 5.1\times 10^{-4}$，平均流速为

$$v = \frac{4\times 1.06}{\pi\times 0.59^2} = 3.87 \text{ (m/s)}$$

and 且

$$Re = \frac{1.35\times 10^6}{0.59} = 2.29\times 10^6$$

The critical Reynolds number for the region of quadratic resistance law is

阻力平方区的极限雷诺数为

$$Re = 4160\left(\frac{1}{2\Delta}\right)^{0.85} = 4160\times\left(\frac{1}{2\times 5.1\times 10^{-4}}\right)^{0.85} = 1.45\times 10^6 < 2.29\times 10^6$$

It can be seen that the flow falls in the region of quadratic resistance law, so λ can be calculated by

可见，流动属阻力平方区，故 λ 可用下式计算

$$\frac{1}{\sqrt{\lambda}} = -2\lg\frac{\Delta}{3.7} = -2\lg\frac{5.1\times 10^{-4}}{3.7}$$

The solution is $\lambda = 0.01677$ and it is unmatched for $\lambda = 0.02$. Now we can assume $\lambda = 0.01677$ and repeat the calculation process

$\lambda = 0.01677$，与假定的 $\lambda = 0.02$ 不符，所以须重新计算。此时假定 $\lambda = 0.01677$，于是得

$$d^5 = 3.56\times 0.01677$$

$$d = 0.569 \text{ (m)}$$

$$\Delta = \delta/d = 0.3\times 10^{-3}/0.569 = 5.27\times 10^{-4}$$

$$v = \frac{4\times 1.06}{\pi\times 0.569^2} = 4.17 \text{ (m/s)}$$

$$Re = \frac{1.35\times 10^6}{d} = \frac{1.35\times 10^6}{0.569} = 2.37\times 10^6 > 1.45\times 10^6$$

The flow still falls in the region of quadratic resistance law, then

流动仍属阻力平方区。则

$$\frac{1}{\sqrt{\lambda}} = -2\lg\frac{\Delta}{3.7} = -2\lg\frac{5.27\times 10^{-4}}{3.7}$$

The solution is $\lambda = 0.0169$. It approaches 0.01677 closely. So d can be determined as 0.569m.

$\lambda = 0.0169$ 近似等于假设值 $\lambda = 0.01677$，于是可取 $d = 0.569$m。

(2) Head loss in series pipes

A serial pipe system is composed of several pipes with different diameters, and the mass flow rate in each pipe is constant, i. e.

(2) 串联管路损失

串联管路由几个不同直径的管段串联在一起组成。各简单管路内质量流量相等，即

$$\dot{m}=\dot{m}_1=\dot{m}_2=\cdots=\rho_1 Q_1=\rho_2 Q_2=\cdots$$

If $\rho_1=\rho_2=\cdots=\rho$, then

若 $\rho_1=\rho_2=\cdots=\rho$，那么

$$Q=Q_1=Q_2=\cdots=v_1 A_1=v_2 A_2=\cdots \quad (7.46)$$

The total head loss $\Sigma \Delta p$ is the summation of the losses in all pipes

串联管路上的总阻力损失等于各支管阻力损失之和，即

$$\Sigma \Delta p=\Delta p_1+\Delta p_2+\cdots=\left(\lambda_1 \frac{l_1}{d_{e1}}+\Sigma \xi_1\right)\frac{\rho_1 v_1^2}{2}+\left(\lambda_2 \frac{l_2}{d_{e2}}+\Sigma \xi_2\right)\frac{\rho_2 v_2^2}{2}+\cdots \quad (7.47)$$

The subsequent steps are similar to that for a single pipe calculation.

接下来的计算步骤与简单管路相似。

(3) Head loss in pipes in parallel

(3) 并联管路损失

A parallel pipeline system is composed of two or more single pipes (or series pipes) by joining their inlets as well as their outlets, as shown in Fig. 7.28.

将两条以上简单管路或串联管路的入口端与出口端分别连接，就组成一个并联管路，如图 7.28 所示。

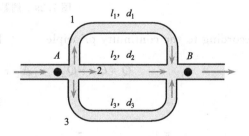

Fig. 7.28   Pipes in parallel
图 7.28   并联管路

According to the mass conservation principle, the flow rate at point A is equal to that at point B, and equal to the summation of the flow rates of all branches. So

根据质量守恒，连接点 A、B 处的质量流量相等，且等于各支管质量流量之和，即

$$\rho Q=\rho_1 Q_1+\rho_2 Q_2+\rho_3 Q_3+\cdots=\rho_1 v_1 A_1+\rho_2 v_2 A_2+\rho_3 v_3 A_3+\cdots \quad (7.48)$$

All the branches have the same head loss, i.e.

$$\Delta p_{AB} = \Delta p_1 = \Delta p_2 = \Delta p_3 = \cdots \quad (7.49)$$

Thereby

$$\left(\lambda_1 \frac{l_1}{d_{e1}} + \Sigma \xi_1\right)\frac{\rho_1 v_1^2}{2} = \left(\lambda_2 \frac{l_2}{d_{e2}} + \Sigma \xi_2\right)\frac{\rho_2 v_2^2}{2} = \left(\lambda_3 \frac{l_3}{d_{e3}} + \Sigma \xi_3\right)\frac{\rho_3 v_3^2}{2} = \cdots \quad (7.50)$$

Using Eq. (7.48) and Eq. (7.49) jointly, the parallel pipeline problem can be solved.

【EXAMPLE 7.4】 An oil flows through the parallel pipeline from point A to point B, as shown in Fig. 7.29, where $l_1 = 30$m, $d_1 = 0.15$m, $l_2 = 400$m, $d_2 = 0.1$m and the flow rate Q is 0.045m³/s. If $\lambda = 0.025$ and minor losses are neglected, find the average velocities in pipes 1 and 2, i.e., v_1 and v_2, and head loss.

并联管路各条支管中流体的阻力损失相等，因此有

从而有

利用式（7.48）和式（7.49）就可对并联管路进行计算。

【例题 7.4】 油料经图 7.29 所示的并联管路由 A 点流向 B 点。体积流量 $Q = 0.045$m³/s，$\lambda = 0.025$，不计次要损失，求支路中的速度 v_1、v_2 以及压头损失。

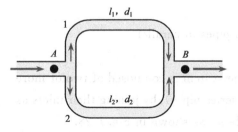

Fig. 7.29 Diagram for example 7.4
图 7.29 例题 7.4 图

SOLUTION According to the continuity principle

解 根据连续性原理

$$Q = Q_1 + Q_2 = v_1 A_1 + v_2 A_2 \quad (7.51)$$

i.e. 即

$$\frac{\pi}{4} \times 0.15^2 v_1 + \frac{\pi}{4} \times 0.1^2 v_2 = 0.045 (\text{m}^3/\text{s}) \quad (7.52)$$

The pressure drops of the two pipes are equal, so

两个支管的压降相等

$$\Delta p_1 = \Delta p_2 = \Delta p_{AB} \quad (7.53)$$

i.e. 即

$$\lambda_1 \frac{l_1}{d_1}\frac{\rho v_1^2}{2} = \lambda_2 \frac{l_2}{d_2}\frac{\rho v_2^2}{2} \quad (7.54)$$

Substituting the given parameters into Eq. (7.54) gives

将已知参数代入式(7.54)，有

$$0.025\frac{300}{0.15}v_1^2 = 0.025\frac{400}{0.1}v_2^2 \quad (7.55)$$

Thereby

从而得

$$v_1 = \sqrt{2}\,v_2 \quad (7.56)$$

Substituting Eq. (7.56) into Eq. (7.52), v_1 and v_2 can be solved as

将式(7.56)代入式(7.52)，可得

$$v_2 = 1.37 \text{ (m/s)}$$

$$v_1 = 1.94 \text{ (m/s)}$$

So the head loss in the parallel pipeline system is

该并联管路的压头损失为

$$h_w = \frac{\Delta p_{AB}}{\rho g} = \lambda_1 \frac{l_1}{d_1}\frac{v_1^2}{2g} = 0.025 \times \frac{300}{0.15} \times \frac{1.94^2}{2 \times 9.81} = 9.6 \text{ (m)}$$

Exercises

7.1 What is meant by equivalent length in relation to pipe system losses?

7.2 Sudden expansion and contractions result in pressure drops across pipe systems. Diffusers are utilized to minimize these losses. Illustrate how diffusers reduce the pressure drop when compared to a sudden expansion.

7.3 Water with a density of 1000kg/m³ is pumped up 50m at a rate of 200 liters per minute. The pipe is 200m long and has an internal diameter of 50mm. The pipeline contains six 90°-elbows and three gate valves. The pump has a 3 kW rating. Using this information calculate: (a) losses due to friction, (b) minor losses, (c) the required pump head, (d) the pump efficiency ($\xi_{\text{gate valve}} = 4.5$, $\xi_{90°\text{-elbow}} = 0.75$).

7.4 It may be a laminar or turbulent flow in smooth pipe when $Re = 3500$. At 20℃ water flows through a

习题

7.1 管路系统中所涉及的当量长度指的是什么？

7.2 管路的突然扩大或缩小会导致系统的压力降，此时会使用扩压器减小损失，请说明扩压器的工作原理。

7.3 密度为 1000kg/m³ 的水以 200L/min 的流量被泵送到50m 的高处。管道长200m，内径50mm。管路上有6个90°弯头和3个闸阀。泵的额定功率为3 kW。计算：(a) 摩擦阻力损失，(b) 次要损失，(c) 泵的最小压头，(d) 泵的效率 ($\xi_{\text{gate valve}} = 4.5$, $\xi_{90°\text{-elbow}} = 0.75$)。

7.4 当 $Re = 3500$ 时，光滑管内的流动可能是层流，也可能是湍流。

smooth pipe with the inner diameter of 50.8mm and the length of 1.3m. Find the ratio of mean velocity, the ratio of head loss and the ratio of wall shear stress under laminar flow condition to that under turbulent condition.

7.5 The air ($\rho = 1.2 \text{kg/m}^3$, $\nu = 1.4 \times 10^{-5} \text{ m}^2/\text{s}$) flows across a pipe with the diameter of 1.25 m and the length of 200m and results in a head loss of 80mmH$_2$O. Calculate the flow rate of air considering $\delta = 1$mm.

7.6 A new castiron pipe without cladding is utilized to transport water at a flow rate of 0.3m^3/s. The head loss is 2 m for a pipe with the length of 1000m. Determine the diameter d of the pipe considering $\nu = 0.87 \times 10^{-6}$ m^2/s.

7.7 Fig. 7.30 illustrates a typical water transport system composed of reservoir, pipe, pump and valve. The pipe inner diameter is 25mm and the flow rate is 0.2m^3/min. The mercury manometer reading and other dimensions can be read in the figure. Find (a) the drag coefficient of the valve, (b) the static pressure right at the upstream of valve, (c) the highest total pressure in the system. The head loss across the pump can be ignored.

7.5 设有 $\rho = 1.2$ kg/m^3、$\nu = 1.4 \times 10^{-5}$ m^2/s 的空气流经直径为 1.25m 的管道，200m 长度内的阻力损失为 80mmH$_2$O，若 $\delta = 1$mm，试计算流量。

7.6 用无镀覆层的新铸铁管来输送流量为 0.3m^3/s 的水，已知在 1000m 长度上的水力损失为 2m，若水的 $\nu = 0.87 \times 10^{-6}$ m^2/s，求管子直径 d。

7.7 图 7.30 所示的输水系统中包括蓄水池、管道、泵和阀。管道内径为 25mm，流量为 0.2m^3/min。水银 U 形压差计及其他参数如图所示。求：(a) 阀门的阻力系数，(b) 阀门前的静压，(c) 系统内的总压。忽略水泵进出口阻力。

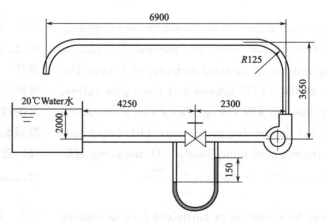

Fig. 7.30 Figure for exercise 7.7
图 7.30 习题 7.7 图

7.8 As shown in Fig. 7.31, the overall flow rate in the serial-parallel pipe system is 0.1 m³/s. Calculate the flow rate in the two branches and the head loss between points A and B. $d_1 = 0.25$m, $l_1 = 1000$m, $d_2 = 0.3$m, $l_2 = 900$m, $d_3 = 0.25$m, $l_3 = 300$m, the roughness of pipe wall is 0.125mm.

7.8 如图 7.31 所示的并联和串联管路，总管中水的流量 0.1m³/s，求各支管流量及 A、B 的压头损失。$d_1 = 0.25$m，$l_1 = 1000$m，$d_2 = 0.3$m，$l_2 = 900$m，$d_3 = 0.25$m，$l_3 = 300$m，管壁的粗糙度为 0.125mm。

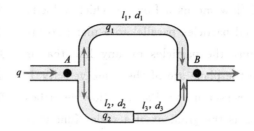

Fig. 7.31 Figure for exercise 7.8
图 7.31 习题 7.8 图

Chapter 8　Planar Potential Flow
第8章　平面势流

Planar potential flow means a flow in which velocity vectors of all fluid particles parallel to a plane at any time; in addition, the particles on any line that is perpendicular to the plane are of the same velocity. It is indeed a two-dimensional flow. Potential flow describes the velocity field as the gradient of a scalar function: the velocity potential. So the planar potential flow introduced in this chapter is a simple two-dimensional potential flow.

Ideal potential flows do not exist because of the boundary influence which is induced by viscous effect and can not be avoided. But applications of potential flow theory are still popular based on some simplifications. For instance: the outer flow field for aerofoils, water waves, electroosmotic flow, and groundwater flow.

平面势流中所有质点在任意时刻的速度都平行于同一固定平面，且垂直于该平面的任意直线上所有流体质点的速度相同，实际上就是二维流动。势流是指速度场可以表示为一个标量函数（速度势函数）梯度的流场。本章介绍的平面势流就是简单的二维势流。

理想的平面势流并不存在，因为实际流场总是存在着由流体黏性引起的边界效应。不过基于简化，某些流动可作为平面势流来处理，例如绕机翼的流动、波浪、电渗流、地下水流等。

8.1　Potential Function and Stream Function

8.1.1　Potential Function

8.1　势函数与流函数

8.1.1　势函数

A velocity potential is a scalar potential used in potential flow theory. It was introduced by Joseph-Louis Lagrangein in 1788 and is valid in continuum mechanics, when a continuum occupies a simply-connected region and is irrotational. So

速度势是势流理论中的一个标量函数，是由拉格朗日在1788年提出的，适用于连续介质力学，要求是单联体且无旋，于是有

$$\frac{\partial v_z}{\partial y}=\frac{\partial v_y}{\partial z},\frac{\partial v_x}{\partial z}=\frac{\partial v_z}{\partial x},\frac{\partial v_y}{\partial x}=\frac{\partial v_x}{\partial y} \qquad (8.1)$$

There exists a velocity potential function $\varphi(x, y, z)$, which satisfies

此时存在一速度势函数 $\varphi(x, y, z)$，满足

$$d\varphi(x,y,z) = v_x dx + v_y dy + v_z dz$$

Equation (8.1) is the necessary and sufficient condition for this total differential equation, and since

式(8.1)是存在此全微分的充分和必要条件,又由于

$$d\varphi(x,y,z) = \frac{\partial \varphi}{\partial x}dx + \frac{\partial \varphi}{\partial y}dy + \frac{\partial \varphi}{\partial z}dz$$

The comparison of the above two equations gives

比较上两式可得

$$v_x = \frac{\partial \varphi}{\partial x}, v_y = \frac{\partial \varphi}{\partial y}, v_z = \frac{\partial \varphi}{\partial z} \quad (8.2)$$

Equation (8.2) illustrates that the velocity projection on each coordinate axis is equal to the partial derivative of velocity potential equation with respect to each coordinate.

式(8.2)说明,速度矢量在三个坐标轴上的投影,等于速度势函数对相应坐标方向的偏导数。

Generally, this characteristic of the velocity potential equation is valid for any direction, i.e., the component of the velocity **v** at a point A on a curve l, v_l (illustrated in Fig. 8.1), is equal to the partial derivative of velocity potential equation with respect to l, i.e., $v_l = \frac{\partial \varphi}{\partial l}$.

速度势函数的这一重要性质,对任何方向来说都是正确的。即任意点 A 上的速度 **v** 沿任一曲线 l 方向的分量 v_l(图8.1),等于该点速度势函数 φ 沿 l 方向的方向导数,$v_l = \frac{\partial \varphi}{\partial l}$。

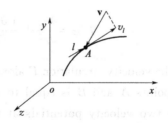

Fig. 8.1 Velocity component with respect to any spatial curve
图 8.1 相对于任意空间曲线的速度分量

Similarly, in two dimensional polar coordinates, we have

类似地,在二维极坐标中有

$$v_r = \frac{\partial \varphi}{\partial r}, v_\theta = \frac{1}{r}\frac{\partial \varphi}{\partial \theta} \quad (8.3)$$

where v_r is the radial velocity component and v_θ the tangential velocity component.

式中,v_r 为径向速度;v_θ 为切向速度。

The characteristics of velocity potential are summarized as the following.

下面分析势函数的一些性质及物理意义。

① The velocity potential of an irrotational flow of incompressible fluid satisfies the Laplace equation. The mass conservation equation for incompressible fluids is in the form

$$\nabla \cdot \mathbf{v} = \frac{\partial v_x}{\partial x} + \frac{\partial v_y}{\partial y} + \frac{\partial v_z}{\partial z} = 0$$

Substitution of Eq. (8.2) into the above equation gives

$$\nabla^2 \varphi = \frac{\partial^2 \varphi}{\partial x^2} + \frac{\partial^2 \varphi}{\partial y^2} + \frac{\partial^2 \varphi}{\partial z^2} = 0 \quad (8.4)$$

This is the Laplace equation in three-dimensional system of coordinate. As for planar potential flow, the Laplace equation reduces to

$$\nabla^2 \varphi = \frac{\partial^2 \varphi}{\partial x^2} + \frac{\partial^2 \varphi}{\partial y^2} = 0 \quad (8.5)$$

In polar system of coordinates, the mass conservation equation holds

$$\frac{v_r}{r} + \frac{\partial v_r}{\partial r} + \frac{1}{r}\frac{\partial v_\theta}{\partial \theta} = 0$$

and the corresponding Laplace equation is

$$\nabla^2 \varphi = \frac{\partial^2 \varphi}{\partial r^2} + \frac{1}{r}\frac{\partial \varphi}{\partial r} + \frac{1}{r^2}\frac{\partial^2 \varphi}{\partial \theta^2} = 0 \quad (8.6)$$

② In potential flow, the velocity circulation Γ along a curve connecting two points A and B is equal to the difference between the two velocity potentials at the two points. It is independent of the shape of the curve and holds

$$\Gamma = \int_A^B (v_x \mathrm{d}x + v_y \mathrm{d}y + v_z \mathrm{d}z)$$

Substitution of Eq. (8.2) in the above equation gives

$$\Gamma = \int_A^B \left(\frac{\partial \varphi}{\partial x}\mathrm{d}x + \frac{\partial \varphi}{\partial y}\mathrm{d}y + \frac{\partial \varphi}{\partial z}\mathrm{d}z\right) = \int_A^B \mathrm{d}\varphi = \varphi_B - \varphi_A \quad (8.7)$$

If point A coincides with point B, the curve is closed, and $\Gamma = 0$. So it is indicated that if the velocity potential is a single valued function, the velocity cir-

① 不可压缩无旋流动的速度势满足拉普拉斯方程。不可压缩流体的连续方程为

将式(8.2)代入上式则得

此式即为三维拉普拉斯方程。对于平面势流,不可压缩流体的连续方程可写成

平面极坐标中不可压缩流体的连续方程为

平面极坐标中的拉普拉斯方程为

② 势流中,沿任一曲线 AB 的速度环量等于 A、B 两点的速度势之差,与曲线形状无关。速度环量为

将式(8.2)代入上式可得

若 A、B 点重合,则 AB 为封闭曲线,$\Gamma = 0$。由此可知:若速度势函数是单值函数,则在无旋流动

culation equals zero for any closed curve in an irrotational flow. But this conclusion is not valid for non-single valued potential function.

③ The velocity component along an arbitrary direction is equal to the partial derivative of velocity potential with respect to the direction.

8.1.2 Stream Function

The stream function is defined for incompressible flows in two-dimensions, as well as in three-dimensions with axisymmetry. The stream function can be used to plot streamlines, which represents the trajectories of particles in a steady flow. According to its definition, the streamline differential equation for planar flow is

$$\frac{\mathrm{d}x}{v_x} = \frac{\mathrm{d}y}{v_y} \text{ or } -v_y \mathrm{d}x + v_x \mathrm{d}y = 0 \quad (8.8a)$$

If $\frac{\partial(-v_y)}{\partial y} = \frac{\partial v_x}{\partial x}$, there exists a total differential equation for Eq. (8.8a). Obviously it is just the mass conservation equation for two-dimensional flows. So there exists a total differential for a function of ψ

$$\mathrm{d}\psi = \frac{\partial \psi}{\partial x}\mathrm{d}x + \frac{\partial \psi}{\partial y}\mathrm{d}y \quad (8.8b)$$

which satisfies both

$$\left.\begin{array}{l} v_x = \dfrac{\partial \psi}{\partial y} \\ v_y = -\dfrac{\partial \psi}{\partial x} \end{array}\right\} \quad (8.9)$$

and Eq. (8.8a). Substitution of Eq. (8.9) into Eq. (8.8b) gives

$$\mathrm{d}\psi = \frac{\partial \psi}{\partial x}\mathrm{d}x + \frac{\partial \psi}{\partial y}\mathrm{d}y = -v_y \mathrm{d}x + v_x \mathrm{d}y = 0 \quad (8.10)$$

Integration of Eq. (8.10) gives

$$\psi(x,y) = C_1 \quad (8.11)$$

中，沿任意封闭曲线的环量为零。对非单值势函数，该结论不成立。

③ 速度在任意方向的分量为势函数对该方向的偏导数。

8.1.2 流函数

二维或三维轴对称不可压缩流动具有流函数。通过流函数可以得到流线，对于稳态流动，流线就是质点的迹线。根据流线的定义，二维流动的流线微分方程为

上式存在全微分的条件是 $\frac{\partial(-v_y)}{\partial y} = \frac{\partial v_x}{\partial x}$，显然该式就是连续性方程，所以存在一个全微分

该全微分满足

以及式(8.8a)。将式(8.9)代入式(8.8b)得到

对式(8.10)积分，就可得到流线方程

where C_1 is a constant. ψ is the so-called stream function. Since C_1 can be valued arbitrarily, there can be a cluster of streamlines.

Comparison of Eq. (8.2) and Eq. (8.9) gives the relationship between the streamline function and velocity potential in the rectangular coordinate system

$$\left. \begin{aligned} v_x &= \frac{\partial \varphi}{\partial x} = \frac{\partial \psi}{\partial y} \\ v_y &= \frac{\partial \varphi}{\partial y} = -\frac{\partial \psi}{\partial x} \end{aligned} \right\} \quad (8.12)$$

In the system of polar coordinates, the relationship is in this form

$$\left. \begin{aligned} v_r &= \frac{\partial \varphi}{\partial r} = \frac{1}{r}\frac{\partial \psi}{\partial \theta} \\ v_\theta &= \frac{1}{r}\frac{\partial \varphi}{\partial \theta} = -\frac{\partial \psi}{\partial r} \end{aligned} \right\} \quad (8.13)$$

Let the velocity potential equal to a constant C_2, we have

$$\varphi(x, y) = C_2 \quad (8.14)$$

Similarly, C_2 can also be valued arbitrarily, resulting in a cluster of equipotential lines.

The characteristics of stream function ψ are summarized as the following.

① In a two-dimensional steady potential flow, streamlines are perpendicular to equipotential lines mutually. The cross-multiplication of the partial derivatives in Eq. (8.12) can be expressed as

$$\frac{\partial \varphi}{\partial x}\frac{\partial \psi}{\partial x} + \frac{\partial \varphi}{\partial y}\frac{\partial \psi}{\partial y} = 0$$

It can be seen that, in the steady planar potential flow, the streamlines and equipotential lines are orthotropic.

② The streamline function for potential flow is a harmonic

function since it satisfies the Laplace equation. For a planar potential flow, $\frac{\partial v_x}{\partial y} - \frac{\partial v_y}{\partial x} = 0$, therefore $\frac{\partial^2 \varphi}{\partial x \partial y} = \frac{\partial^2 \varphi}{\partial y \partial x}$. Considering the relationship between φ and ψ [Eq. (8.12)], we have $\frac{\partial^2 \psi}{\partial x^2} + \frac{\partial^2 \psi}{\partial y^2} = 0$.

满足拉普拉斯方程式。由 $\frac{\partial v_x}{\partial y} - \frac{\partial v_y}{\partial x} = 0$ 得 $\frac{\partial^2 \varphi}{\partial x \partial y} = \frac{\partial^2 \varphi}{\partial y \partial x}$。再利用式(8.12) 得 $\frac{\partial^2 \psi}{\partial x^2} + \frac{\partial^2 \psi}{\partial y^2} = 0$。

③ The difference between the stream function values at any two points gives the volumetric flow rate (or volumetric flux) through a line connecting the two points, which is independent of the shape of the curve.

③ 流经任意曲线的体积流量（或体积通量）等于曲线两端点上流函数值之差，而与曲线形状无关。

As shown in Fig. 8.2, **n** is the exterior normal of point M on line AB in a planar potential flow. The flow rate through the area about point M and with the length of dl and the unit width of 1 is

图 8.2 所示的平面势流中，M 为曲线 AB 上任一点，**n** 为外法线。通过点 M 附近长为 dl、宽为 1 的微元面的流量为

$$dQ = v_n dl$$

where v_n is the velocity component with respect to **n**

式中，v_n 为沿 **n** 的速度分量

$$v_n = v_x \cos(\mathbf{n},\mathbf{i}) + v_y \cos(\mathbf{n},\mathbf{j})$$

Thereby

因此

$$dQ = [v_x \cos(\mathbf{n},\mathbf{i}) + v_y \cos(\mathbf{n},\mathbf{j})]dl$$

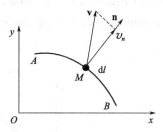

Fig. 8.2　Velocity component with respect to normal direction **n** of any special curve

图 8.2　相对于空间曲线上某点法线方向的速度方向

Considering

考虑到

$$\cos(\mathbf{n},\mathbf{i}) = \frac{dy}{dl}, \cos(\mathbf{n},\mathbf{j}) = -\frac{dx}{dl}$$

dQ can be expressed as

dQ 可表述为

$$dQ = \left[\frac{\partial \psi}{\partial y}\frac{dy}{dl} + \left(-\frac{\partial \psi}{\partial x}\right)\left(-\frac{dx}{dl}\right)\right]dl = \frac{\partial \psi}{\partial x}dx + \frac{\partial \psi}{\partial y}dy = d\psi \qquad (8.15)$$

Equation (8.15) indicates that the differential of stream function equals the flow rate through the representative area with length of dl and width of 1. Therefore the total flow rate across curve AB is

式(8.15)表明,流函数的微分就等于单位时间内流经长为 dl、宽为 1 的微元曲面的流量。因此流过曲线 AB 的流量为

$$Q = \int_A^B d\psi = \psi_B - \psi_A \qquad (8.16)$$

From Eq. (8.16) we can understand that if line AB itself coincidently presents a streamline, the flow rate Q is zero because ψ is constant along a streamline. If points A and B locate on different streamlines ($\psi = C_A$ and $\psi = C_B$), the flow rate across line AB is a constant equal to $Q = \psi_A - \psi_B = C_A - C_B$ and independent of the positions of the two points. If point A coincides with point B, line AB is closed and the flow rate $Q=0$ if ψ is a single-valued function; if ψ is not a single-valued function, Q is not zero.

式(8.16)表明,若曲线 AB 本身就是流线,因沿流线 $\psi=$ 常数,故流过 AB 曲线的流量 $Q=0$。若 A、B 落在不同的流线上($\psi=C_A$ 和 $\psi=C_B$),则流经曲线 AB 的流量是定值,为 $Q=\psi_A-\psi_B=C_A-C_B$,与 A、B 点的位置无关。若 A 和 B 重合,曲线 AB 封闭,若 ψ 为单值,$Q=0$;若 ψ 非单值函数,则 Q 不等于零。

8.2 Simple Potential Flow

8.2.1 Uniform Linear Flow

The streamlines and equipotential lines of a uniform linear flow are illustrated in Fig. 8.3. In this case, all the particles move at the same velocity along the same direction.

8.2 简单势流

8.2.1 均匀直线流动

图 8.3 所示为均匀直线流动的流线和等势线。该流场中各点的速度大小相等、方向相同。

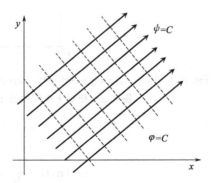

Fig. 8.3 Uniform linear flow

图 8.3 均匀直线流动

The velocity vector **v** is inclined at the angle, α, with respect to the x axis, so the velocity components can be expressed as

$$\left.\begin{array}{l} v_x = v\cos\alpha \\ v_y = v\sin\alpha \end{array}\right\} \quad (8.17\text{a})$$

and the velocity potential and streamline functions can be obtained by

$$\left.\begin{array}{l} \mathrm{d}\varphi = v_x \mathrm{d}x + v_y \mathrm{d}y = v\cos\alpha \mathrm{d}x + v\sin\alpha \mathrm{d}y \\ \mathrm{d}\psi = v_x \mathrm{d}y - v_y \mathrm{d}x = v\cos\alpha \mathrm{d}y - v\sin\alpha \mathrm{d}x \end{array}\right\} \quad (8.17\text{b})$$

Integration of the equations in Eq. (8.17b) gives

$$\left.\begin{array}{l} \varphi = v(x\cos\alpha + y\sin\alpha) = C_1 \\ \psi = v(y\cos\alpha - x\sin\alpha) = C_2 \end{array}\right\} \quad (8.18)$$

Obviously

$$\frac{\partial \varphi}{\partial x}\frac{\partial \psi}{\partial x} + \frac{\partial \varphi}{\partial y}\frac{\partial \psi}{\partial y} = v\cos\alpha(-v\sin\alpha) + v\sin\alpha \cdot v\cos\alpha = 0$$

So it can be seen that the equipotential lines are perpendicular to the streamlines. If the flow is parallel to the x axis, φ and ψ are in the form

$$\left.\begin{array}{l} \varphi = vx = C_1 \\ \psi = vy = C_2 \end{array}\right\} \quad (8.19)$$

and if the flow is parallel to the y axis, they are in the form

$$\left.\begin{array}{l} \varphi = vy = C_1 \\ \psi = -vx = C_2 \end{array}\right\} \quad (8.20)$$

8.2.2 Flow in Right-angle Region

The velocity potential equation of planar potential flow in right-angle region is

$$\varphi = a(x^2 - y^2) \quad (8.21)$$

where a is a real number and $a > 0$. So the corresponding streamline function ψ can be determined according to φ. According to Eq. (8.12)

$$\psi = \int \frac{\partial \varphi}{\partial x} dy + C(x) \qquad (8.22a)$$

where C is a function of x. Substituting Eq. (8.21) into Eq. (8.22a) and then integrating, Eq. (8.22b) can be obtained.

式中，常数 C 是 x 的函数。将式 (8.21) 代入式 (8.22a) 并积分，可得

$$\psi = \int 2ax \, dy + C(x) = 2axy + C(x) \qquad (8.22b)$$

To determine the function $C(x)$, we differentiate Eq. (8.22b) with respect to x

为了确定 $C(x)$，将式 (8.22b) 对 x 求导得

$$\frac{\partial \psi}{\partial x} = 2ay + \frac{dC(x)}{dx} \qquad (8.22c)$$

Considering the relationship shown by Eq. (8.12), Eq. (8.22c) can be rewritten as

根据式 (8.12) 的关系，式 (8.22c) 可写成

$$-\frac{\partial \varphi}{\partial y} = 2ay + \frac{dC(x)}{dx}$$

Based on Eq. (8.21), $\frac{\partial \varphi}{\partial y} = -2ay$, therefore $\frac{dC(x)}{dx} = 0$, which means $C(x) = C_1$. Thereby the stream function holds $\psi = 2axy + C_1$ and the streamline function is

又由式 (8.21) 可得 $\frac{\partial \varphi}{\partial y} = -2ay$，即 $C(x) = C_1$。得到流函数为 $\psi = 2axy + C_1$，由此得流线方程为

$$xy = C \qquad (8.23)$$

Equation (8.23) indicates that the equipotential lines are a cluster of hyperbolic curves, as shown in Fig. 8.4.

由式 (8.23) 可知，等势线亦是一簇双曲线，如图 8.4 所示。

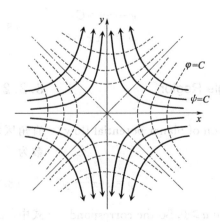

Fig. 8.4　Streamlines and equipotential lines of planar potential flow in right-angle regions
图 8.4　直角区域内平面势流的流线和等势线

8.2.3　Point Source and Point Sink

Fluid point sources are commonly encountered in fluid dynamics and aerodynamics. A point source of fluid is the inverse of a fluid point sink (a point where fluid is removed). Whereas fluid sinks exhibit complex rapidly changing behavior such as is seen in vortices (for example water running into a plug-hole or tornadoes generated at points where air is rising), fluid sources generally produce simple flow patterns, with stationary isotropic point sources generating an expanding sphere of new fluid. If the fluid is moving (such as wind in air or currents in water) a plume is generated from the point source.

Figure 8.5 illustrates the streamlines and equipotential lines of a point source, where point o is the source point.

8.2.3　点源和点汇

在流体动力学和空气动力学中会经常遇到点源流动。点源是点汇的反过程。点汇具有更复杂的变化特性，会出现旋涡（如水流入放水孔或空气上升时形成的龙卷风）；而点源流动的流型较简单，静止的各向同性的点源生成了不断扩大的圆。如果流体处于运动状态（如风或水流），会从点源处生成烟羽。

图 8.5 所示为点源的流线和等势线，o 点称为源点。

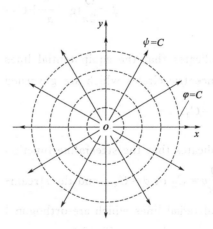

Fig. 8.5　Point source
图 8.5　点源

In the polar coordinate system, as illustrated in Fig. 8.5, the volume flow rate from the original point is Q, so at a point $P(r, \theta)$, the radial velocity is

$$v_r = \frac{Q}{2\pi r} \quad (8.24a)$$

The velocity is along the radial direction, r, and the velocity components along the x and y axis are

以源点 o 为坐标原点取极坐标系，若其体积流量为 Q，则任意点 $P(r, \theta)$ 的流速为

速度方向与半径 r 方向一致，其 x、y 方向的速度分量为

$$\left. \begin{array}{l} v_x = v_r \cos\theta = \dfrac{Q}{2\pi r}\dfrac{x}{r} = \dfrac{Q}{2\pi}\dfrac{x}{x^2+y^2} \\ v_y = v_r \sin\theta = \dfrac{Q}{2\pi r}\dfrac{y}{r} = \dfrac{Q}{2\pi}\dfrac{y}{x^2+y^2} \end{array} \right\} \quad (8.24b)$$

Therefore

$$\mathrm{d}\varphi = v_x \mathrm{d}x + v_y \mathrm{d}y = \frac{Q}{2\pi}\frac{x\mathrm{d}x + y\mathrm{d}y}{x^2+y^2} = \frac{Q}{4\pi}\frac{\mathrm{d}(x^2+y^2)}{x^2+y^2}$$

and the velocity potential equation is 速度势函数为

$$\varphi = \frac{Q}{4\pi}\ln(x^2+y^2) + C_1 = \frac{Q}{2\pi}\ln r + C_1 \quad (8.25)$$

Furthermore 因此

$$\mathrm{d}\psi = v_x \mathrm{d}y - v_y \mathrm{d}x = \frac{Q}{2\pi}\frac{x\mathrm{d}y - y\mathrm{d}x}{x^2+y^2} = \frac{Q}{2\pi}\frac{\mathrm{d}\left(\dfrac{y}{x}\right)}{1+\left(\dfrac{y}{x}\right)^2}$$

$$\psi = \frac{Q}{2\pi}\mathrm{tg}^{-1}\frac{y}{x} + C_2 = \frac{Q}{2\pi}\theta + C_2 \quad (8.26)$$

Equation(8.25) indicates that the equipotential lines are a cluster of concentric circles which are governed by $\dfrac{Q}{2\pi}\ln r = C_1$ or $r = C_1$.

式(8.25) 表明，等势线方程为 $\dfrac{Q}{2\pi}\ln r = C_1$ 或 $r = C_1$，即等势线是以源点 o 为圆心的同心圆。

Equation(8.26) indicates that the streamline function is in the form of $\dfrac{Q}{2\pi}\theta = C_2$ or $\theta = C_2$, and the streamlines are a cluster of radial lines which are orthogonal to equipotential lines, as shown in Fig. 8.5.

由式(8.26) 可知流线方程为 $\dfrac{Q}{2\pi}\theta = C_2$ 或 $\theta = C_2$，它是以源点 o 为起点的半辐射线，与等势线正交，如图 8.5 所示。

Since the point sink (Fig. 8.6) is a reverse course of point source, the velocity potential equation and the streamline function can be easily obtained from Eq. (8.25) and Eq. (8.26), i.e.

点汇流动是点源流动的逆过程，如图 8.6 所示，其势函数和流函数可从式(8.25) 和式(8.26) 简单变化得到

$$\left. \begin{array}{l} \varphi = -\dfrac{Q}{2\pi}\ln r \\ \psi = -\dfrac{Q}{2\pi}\theta \end{array} \right\} \quad (8.27)$$

From Eq. (8.24a), it can be seen that for both point source and point sink the theoretical velocity at the original point o is infinite, so the velocity is not continuous at this point. The flow rate across a closed curve which besieges point o is always equal to Q.

由式(8.24a)可知，点源和点汇在原点 o 处流速无穷大，故原点处速度不连续，且流经包围源点或汇点的任何封闭曲线上的流量都等于 Q。

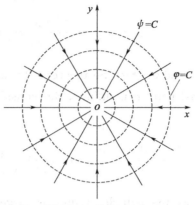

Fig. 8.6 Point sink
图 8.6 点汇

8.2.4 Pure Circulation Flow

As illustrated in Fig. 8.7, a cylinder with the radius of r_0 is rotating about its axis at the speed of ω. Due to sticking effect, the fluid clinging to the cylinder is drawn to rotate around the cylinder. The velocity at point $P(r, \theta)$ is along the circumferential direction and it is inversely proportional to r, i.e., $v = K/r$ where K is a constant. As r trends to infinity, v trends to be zero. This type of flow is termed the pure circulation flow.

8.2.4 纯环流流动

如图 8.7 所示，半径为 r_0 圆柱体绕轴以角速度 ω 旋转，周围流体将被带动跟着做旋转运动。点 $P(r, \theta)$ 的速度方向为圆周方向，其大小与 r 成反比，即 $v = K/r$，K 是常数。当 r 趋近于 ∞ 时，v 趋近于 0。这种平面流动称为纯环流流动。

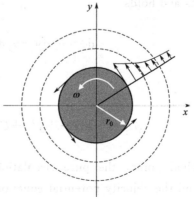

Fig. 8.7 Pure circulation flow
图 8.7 纯环流流动

On the cylinder surface ($r=r_0$), $v=r_0\omega$. Thereby $K=r_0^2\omega$ and $v=r_0^2\omega/r$. The velocity components along the x and y directions are

$$\left.\begin{aligned}v_x&=-v\sin\theta=-\frac{r_0^2\omega}{r}\frac{y}{r}=-r_0^2\omega\frac{y}{x^2+y^2}\\ v_y&=v\cos\theta=\frac{r_0^2\omega}{r}\frac{x}{r}=r_0^2\omega\frac{x}{x^2+y^2}\end{aligned}\right\} \quad (8.28)$$

When $r\to 0$, $v_x\to\infty$ and $v_y\to\infty$. So theoretically the velocity at the original point o is infinite. But considering

$$\frac{\partial v_x}{\partial x}+\frac{\partial v_y}{\partial y}=\frac{\partial}{\partial x}\left(-r_0^2\omega\frac{y}{x^2+y^2}\right)+\frac{\partial}{\partial y}\left(r_0^2\omega\frac{x}{x^2+y^2}\right)=\frac{0}{(x^2+y^2)^2}=0$$

Therefore, so long as the cylinder radius is larger than zero, the pure circulation flow is continuous, and the velocity circulation across any closed curve that besieges the original point is constant, which is

$$\Gamma=\oint_l v\cdot\mathrm{d}l=\int_0^{2\pi}\frac{r_0^2\omega}{r}r\mathrm{d}\theta=2\pi r_0^2\omega \quad (8.29)$$

Then Eq. (8.28) can be rewritten as

$$\left.\begin{aligned}v_x&=-\frac{\Gamma}{2\pi}\frac{y}{x^2+y^2}\\ v_y&=\frac{\Gamma}{2\pi}\frac{x}{x^2+y^2}\end{aligned}\right\} \quad (8.30)$$

Except for the original point, the stream function for pure circulation exists and holds

$$\mathrm{d}\psi=v_x\mathrm{d}y-v_y\mathrm{d}x=-\frac{\Gamma}{2\pi}\frac{y\mathrm{d}y+x\mathrm{d}x}{x^2+y^2}$$

By integrating

$$\psi=-\frac{\Gamma}{4\pi}\ln(x^2+y^2)+C_1=-\frac{\Gamma}{2\pi}\ln r+C_1 \quad (8.31)$$

Except for the original point, the pure circulation flow is irrotational and the velocity potential equation can be obtained from

$$d\varphi = v_x dx + v_y dy = \frac{\Gamma}{2\pi} \frac{-y dx + x dy}{x^2 + y^2} = \frac{\Gamma}{2\pi} \frac{d\left(\frac{y}{x}\right)}{1 + \left(\frac{y}{x}\right)^2}$$

By integrating 对上式积分

$$\varphi = \frac{\Gamma}{2\pi} \mathrm{tg}^{-1} \frac{y}{x} + C_2 = \frac{\Gamma}{2\pi} \theta + C_2 \qquad (8.32)$$

From Eq. (8.31) and Eq. (8.32) it can be seen that the streamlines of pure circulation flow are a cluster of concentric circles while the equipotential lines are a cluster of half-lines with respect to the start point of o, as shown in Fig. 8.8.

由式（8.31）和式（8.32）可知纯环流流动的流线是以原点为中心的同心圆，其等势线则是以原点为起点的半辐射线，如图 8.8 所示。

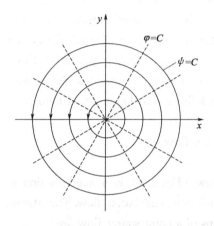

Fig. 8.8 Streamlines and equipotential lines of pure circulation flow
图 8.8 纯环流流动的流线和等势线

8.3 Superposition Principle of Potential Flows

8.3 势流叠加原理

There are two simple potential flows with the velocity potential of φ_1 and φ_2, and they satisfy the Laplace equation

设有两个势流，速度势为 φ_1 和 φ_2，分别满足拉普拉斯方程

$$\left. \begin{array}{l} \nabla^2 \varphi_1 = 0 \\ \nabla^2 \varphi_2 = 0 \end{array} \right\} \qquad (8.33)$$

Combining the two equations gives 联立两式，得

$$\nabla^2 (\varphi_1 + \varphi_2) = 0 \qquad (8.34)$$

It can be seen that the combination of two velocity potentials, $\varphi = \varphi_1 + \varphi_2$, presents a new planar potential

可见两个速度势之和也满足拉普拉斯方程，表明速度势 φ（$\varphi = \varphi_1 + \varphi_2$）

flow which also satisfy the Laplace equation.　　可代表某一新的平面势流。

The velocity components of the new flow are　　复合流动的速度分量为

$$\left.\begin{array}{l}v_x=\dfrac{\partial\varphi}{\partial x}=\dfrac{\partial\varphi_1}{\partial x}+\dfrac{\partial\varphi_2}{\partial x}=v_{x1}+v_{x2}\\ v_y=\dfrac{\partial\varphi}{\partial y}=\dfrac{\partial\varphi_1}{\partial y}+\dfrac{\partial\varphi_2}{\partial y}=v_{y1}+v_{y2}\end{array}\right\} \quad (8.35)$$

Similarly, the stream function of a superposition flow is the algebraic sum of the stream functions of two simple flows　　类似的，复合流动的流函数等于两个原始流动流函数的代数和

$$\psi=\psi_1+\psi_2 \quad (8.36)$$

In summary, a new potential flow can be formed by superimposing some simple potential flows together. Correspondingly, for a complex potential flow, we can decompose it into some simple potential flows with known stream and potential functions. Some typical complex potential flows will be analyzed below.　　总的来说，几个简单势流可叠加得到新的势流。同样，在复杂势流问题下，亦可把复杂流动分解成几个简单的、已知流函数和势函数的流动。下面介绍几个典型的复杂势流。

(1) Source circulation flow　　（1）源环流动

Source circulation flow (Fig. 8.9) is a superposition of point source flow and pure circulation flow. The stream and potential functions of a point source flow are　　源环流动（图 8.9）是点源流动与纯环流动的叠加。点源流动的流函数及势函数为

$$\left.\begin{array}{l}\varphi_1=\dfrac{Q}{2\pi}\ln r\\ \psi_1=\dfrac{Q}{2\pi}\theta\end{array}\right\} \quad (8.37)$$

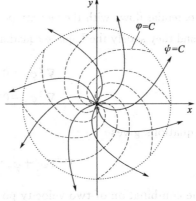

Fig. 8.9　Source circulation flow

图 8.9　源环流动

and for a pure circulation flow 　　　　　　　　对于纯环流动

$$\left.\begin{array}{c}\varphi_2=\dfrac{\Gamma}{2\pi}\theta\\ \psi_2=-\dfrac{\Gamma}{2\pi}\ln r\end{array}\right\} \quad (8.38)$$

So the stream and velocity potential functions of the source circulation flow be expressed as 　　源环流动的流函数和速度势函数可表示为

$$\left.\begin{array}{c}\varphi=\varphi_1+\varphi_2=\dfrac{Q}{2\pi}\ln r+\dfrac{\Gamma}{2\pi}\theta\\ \psi=\psi_1+\psi_2=\dfrac{Q}{2\pi}\theta-\dfrac{\Gamma}{2\pi}\ln r\end{array}\right\} \quad (8.39)$$

The velocity potential function is 　　　　　　　速度势函数为

$$Q\ln r+\Gamma\theta=C_1$$

$$r=e^{\frac{C_1-\Gamma\theta}{Q}} \quad (8.40)$$

and the streamline equation is 　　　　　　　流线方程为

$$Q\theta-\Gamma\ln r=C_2$$

$$r=e^{\frac{C_2+Q\theta}{\Gamma}} \quad (8.41)$$

For a source circulation flow, the equipotential lines are a cluster of log spirals which are orthogonal to the log spiral streamlines, as illustrated in Fig. 8.9.

源环流动的等势线是一簇与螺旋流线正交的对数螺旋线，如图 8.9 所示。

In a centrifugal pump, the flow is governed by Eq. (8.39). If the impeller is static while the water is still supplied, the flow can be attributed to a point source flow; if the impeller is rotating while the water is not supplied, the flow is attributed to a pure circulation flow; if the impeller is rotating and simultaneously the water is supplied, it is a superimposing of a point source flow and a pure circulation flow. Therefore, to avoid the collision between water and impeller, blades are usually streamlined, as shown in Fig. 8.10.

离心泵内流体的流动符合式 (8.39) 所表示的流动规律。当叶轮不转动而供水时，叶轮内的流动为点源流动；当叶轮转动而不供水时，叶轮内的流动为纯环流动；当叶轮转动且同时供水，叶轮内的流动为点源流动与纯环流流动的叠加。为了避免水和叶轮的撞击，叶片通常是流线型的，如图 8.10 所示。

Fig. 8.10 Centrifugal impeller
图 8.10 离心泵的叶轮

(2) Sink circulation flow

A sink circulation flow is the superposition of point sink flow and pure circulation flow. Its stream and velocity potential functions are

（2）汇环流动

汇环流动是点汇流动与纯环流流动叠加的结果。其流函数与势函数为

$$\left.\begin{array}{l}\varphi=\varphi_1+\varphi_2=-\dfrac{Q}{2\pi}\ln r+\dfrac{\Gamma}{2\pi}\theta \\ \psi=\psi_1+\psi_2=-\dfrac{Q}{2\pi}\theta-\dfrac{\Gamma}{2\pi}\ln r\end{array}\right\} \quad (8.42)$$

with the streamline equation

流线方程为

$$r=e^{-\frac{C+Q\theta}{\Gamma}} \quad (8.43)$$

The sink circulation flow is similar to source circulation flow. In the former, fluid flows toward center while in the later, fluid flows from center point to periphery. The sink circulation alike flows can be seen in the air swirl combustion chamber or cyclone dust extractor, etc.

这种流动的特点与源环流动很接近，只是前者是由四周向中心流，后者是由中心向外流。如旋风燃烧室、旋风除尘器等设备中的旋转气流就是一种汇环流动。

(3) Doublet flow

A doublet flow is formed by infinitely approaching a point source flow with flow rate Q to a point sink flow with flow rate $-Q$, as illustrated in Fig. 8.11.

（3）偶极流

使点源无限接近于点汇（其流量各为 Q 和 $-Q$）就形成了偶极流，如图 8.11 所示。

As shown in Fig. 8.11, the distance between the source point and sink point is 2ε. A point on the x-o-y plane, $P(x, y)$, is of r_1 and r_2 distances away from the two points, respectively. So the potential function at the point P is

设源点与汇点间距离为 2ε，平面内任意点 $P(x, y)$ 至点源和点汇的距离分别为 r_1 和 r_2，如图 8.11 所示，则点 P 处的势函数为

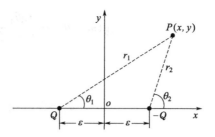

Fig. 8.11 Doublet flow
图 8.11 偶极流

$$\varphi = \varphi_1 + \varphi_2 = \frac{Q}{2\pi}(\ln r_1 - \ln r_2)$$

$$\varphi = \frac{Q}{2\pi}\ln\frac{r_1}{r_2} \quad (8.44a)$$

The distances r_1 and r_2 can be calculated by r_1 和 r_2 可根据下式计算

$$\left.\begin{array}{l} r_1 = \sqrt{(x+\varepsilon)^2 + y^2} \\ r_2 = \sqrt{(x-\varepsilon)^2 + y^2} \end{array}\right\} \quad (8.44b)$$

Substitution of Eq. (8.44b) into Eq. (8.44a) gives 将式(8.44b) 代入式(8.44a)，得

$$\varphi = \frac{Q}{2\pi}\ln\sqrt{\frac{(x+\varepsilon)^2 + y^2}{(x-\varepsilon)^2 + y^2}} = \frac{Q}{4\pi}\ln\frac{(x+\varepsilon)^2 + y^2}{(x-\varepsilon)^2 + y^2} = \frac{Q}{4\pi}\ln\left[1 + \frac{4x\varepsilon}{(x-\varepsilon)^2 + y^2}\right] \quad (8.44c)$$

Let the source point and sink point be infinitely close, i.e., $\varepsilon \to 0$, Eq. (8.44c) can be expanded in series as 若使源点和汇点无限接近，即 $\varepsilon \to 0$，将式(8.44c) 按级数型式展开

$$\ln(1+Z) = Z - \frac{Z^2}{2} + \frac{Z^3}{3} - \cdots$$

and by omitting the higher order terms, Eq. (8.44c) is reduced as 略去高阶微量，式(8.44c) 简化为

$$\varphi = \frac{Q}{4\pi}\frac{4x\varepsilon}{(x-\varepsilon)^2 + y^2} \quad (8.44d)$$

The stream function is P 处的流函数为

$$\psi = \psi_1 + \psi_2 = \frac{Q}{2\pi}(\theta_1 - \theta_2) \quad (8.44e)$$

Considering 由

$$\mathrm{tg}\theta_1 = \frac{y}{x+\varepsilon}, \mathrm{tg}\theta_2 = \frac{y}{x-\varepsilon}$$

and 有

$$\operatorname{tg}(\theta_1-\theta_2)=\frac{\operatorname{tg}\theta_1-\operatorname{tg}\theta_2}{1+\operatorname{tg}\theta_1\operatorname{tg}\theta_2}=\frac{-2y\varepsilon}{x^2+y^2-\varepsilon^2}$$

So 因此

$$\theta_1-\theta_2=\operatorname{tg}^{-1}\frac{-2y\varepsilon}{x^2+y^2-\varepsilon^2}$$

Equation (8.44e) is rewritten as 式(8.44e) 可写为

$$\psi=\frac{Q}{2\pi}\operatorname{tg}^{-1}\frac{-2y\varepsilon}{x^2+y^2-\varepsilon^2}$$

Using the expanded formula of 采用展开式

$$\operatorname{tg}^{-1}Z=Z-\frac{Z^3}{3}+\frac{Z^5}{5}-\cdots$$

and when $\varepsilon\to 0$, Eq. (8.44e) becomes 当 $\varepsilon\to 0$ 时，式(8.44e) 变为

$$\psi=-\frac{Q}{2\pi}\frac{2y\varepsilon}{x^2+y^2-\varepsilon^2} \quad (8.44f)$$

Since $Q\to\infty$ when $\varepsilon\to 0$, $2Q\varepsilon$ trends to the extremum M which is termed the doublet moment of doublet flow. Substituting M into Eq. (8.44d) and Eq. (8.44f), respectively, and let $\varepsilon\to 0$, the velocity potential and stream functions can be obtained as

假定 $\varepsilon\to 0$ 时 $Q\to\infty$，则 $2Q\varepsilon$ 趋于极限值 M (M 称为偶极流的偶极矩)。把 M 值代入式(8.44d) 和式(8.44f)，并使 $\varepsilon\to 0$，则得到偶极流的势函数和流函数为

$$\varphi=\frac{M}{2\pi}\frac{x}{x^2+y^2} \quad (8.45)$$

$$\psi=-\frac{M}{2\pi}\frac{y}{x^2+y^2} \quad (8.46)$$

The streamline equation holds 流线方程为

$$\frac{y}{x^2+y^2}=C_1$$

or 或

$$x^2+\left(y-\frac{1}{2C_1}\right)^2=\frac{1}{4C_1^2}$$

The streamlines of a doublet flow are a cluster of circles with the center of $(0, 0.5C_1)$ and radius of $0.5C_1$ as can be seen in Fig. 8.12, they are tangential to the x-axis. Fluid runs out from the original point and goes back to the original point.

偶极流的流线是圆心为 $(0, 0.5C_1)$、半径为 $0.5C_1$ 的圆簇(图 8.12)，所有圆均与 x 轴相切。流体由坐标原点流出，又流回原点。

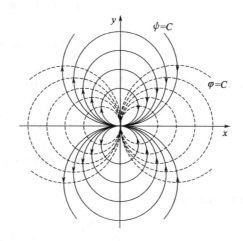

Fig. 8.12 Streamlines and equipotential lines of doublet flow

图 8.12 偶极流的流线和等势线

The equipotential line equation is

等势线方程为

$$\left(x - \frac{1}{2C_2}\right)^2 + y^2 = \frac{1}{4C_2^2}$$

The equipotential lines are a cluster of circles with the centre of $(0.5C_2, 0)$ and radius of $0.5C_2$, which are tangential to the y-axis and shown by the dotted circles in Fig. 8.12.

等势线是圆心为 $(0.5C_2, 0)$、半径为 $0.5C_2$ 的圆簇，如图 8.12 中的虚线所示。

(4) Superimposing of point source flow and uniform linear flow

(4) 平行直线流动与点源的叠加

A point source is located at the original point of the coordinate system, and it encounters a uniform linear flow with the velocity of v_∞ paralleling to the x-axis. According to Eq. (8.19), the velocity potential function, φ, and stream function, ψ, for the uniform linear flow are

将坐标原点取在点源处，平行直线流动的速度为 v_∞，方向自左向右平行于 x 轴。由式 (8.19) 可知，平行直线流动的势函数 φ 和流函数 ψ 为

$$\left.\begin{array}{l}\varphi = v_\infty x \\ \psi = v_\infty y\end{array}\right\}$$

By associating Eq. (8.25) and Eq. (8.26), the velocity potential function, φ, and stream function, ψ, for this complex flow are obtained as

结合式 (8.25) 和式 (8.26)，得到新的复合流动的势函数 φ 和流函数 ψ 分别为

$$\left.\begin{aligned}\varphi &= v_\infty r\cos\theta + \frac{Q}{2\pi}\ln r\\ \psi &= v_\infty r\sin\theta + \frac{Q}{2\pi}\theta\end{aligned}\right\}$$

and the equipotential line and streamline functions are 等势线和流线方程为

$$\left.\begin{aligned}v_\infty r\cos\theta + \frac{Q}{2\pi}\ln r &= C_1\\ v_\infty r\sin\theta + \frac{Q}{2\pi}\theta &= C_2\end{aligned}\right\} \quad (8.47)$$

The velocity components are 速度分量为

$$\left.\begin{aligned}v_r &= \frac{\partial\varphi}{\partial r} = v_\infty\cos\theta + \frac{Q}{2\pi r}\\ v_\theta &= \frac{1}{r}\frac{\partial\varphi}{\partial\theta} = -v_\infty\sin\theta\end{aligned}\right\} \quad (8.48)$$

$$v = \sqrt{v_r^2 + v_\theta^2} = \sqrt{v_\infty^2 + \frac{v_\infty Q\cos\theta}{\pi r} + \frac{Q^2}{4\pi^2 r^2}} \quad (8.49)$$

It can be seen when $r\to 0$, $v\to\infty$ and when $r\to\infty$, $v = v_\infty$. 当 $r\to 0$ 时，$v\to\infty$；当 $r\to\infty$ 时，$v\to v_\infty$。

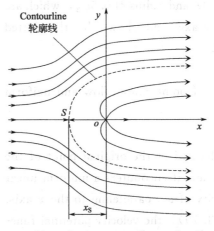

Fig. 8.13 Superposition of point source flow and uniform linear flow
图 8.13 点源和均匀直线流动的叠加

In this complex flow, the velocity near the source point is very large and it decreases with the increasing distance from the source point, as shown in Fig. 8.13. On the negative semi x-axis, the velocity direction of the uniform flow is on the contrary to that of point source flow, so at a point S, the resultant velocity is zero. The distance between points S and o can be calculated by

如图 8.13 所示，在这种复合流动中，接近点源时的流速很大，但流速随离点源距离的增大而减小。由于有不变的平行于 x 轴的流速 v_∞，因此在 x 轴负方向上某点 S 的流速将为 0。此点 S 离原点 o 的距离为

$$x_S = \frac{Q}{2\pi v_\infty} \quad (8.50)$$

Thereby the uniform linear flow will separate at point S and flows around the region denoted by the dashed line shown in Fig. 8.13. The dashed line is just the streamline that passes through point S. The coordinates of points S are

$$\theta = \pi, r = x_S = \frac{Q}{2\pi v_\infty}$$

Therefore

$$\psi|_{\theta=\pi} = v_\infty r \sin\pi + \frac{Q}{2\pi}\pi = \frac{Q}{2}$$

The dashed contourline equation is

$$v_\infty r \sin\theta + \frac{Q\theta}{2\pi} = \frac{Q}{2} \quad (8.51)$$

when $\theta = \frac{\pi}{2}$, $r = y = \frac{Q}{4v_\infty}$ and when $\theta = 0$, $r \to \infty$, but $r\sin\theta = \frac{Q}{2v_\infty}$.

The contourline formed in this kind of complex flow is like a semi body with a head but without a tail. It can be used to investigate the upstream flow around an axisymmetric object, such as pier, pillar, torpedo, and airframe, etc.

Exercises

8.1 Illustrate the existence conditions of potential function and stream function, respectively.

8.2 How a flow can be characterized as an irrotational flow? What are the characteristics of an irrotational flow?

8.3 Establish the potential function and stream function for an incompressible flow with the velocity components of $v_x = 3$m/s and $v_y = 5$m/s.

8.4 Here is a flow field with $v_x = kx^2$ and $v_y = -2kxy$ where k is a constant. Determine whether the velocity components can satisfy the continuity equation and whether it is an irrotaional flow or not. Give the stream function, potential function and streamline equation if existent.

8.5 There is a planar potential flow with the velocity potential of $\varphi = K\theta$ (θ is the polar angle and K is a positive constant). Try to calculate the circulation across the circles of $x^2 + y^2 = R^2$ and $(x-a)^2 + y^2 = R^2$, respectively.

8.6 The fluid flows around an axis at the angular velocity ω which is inversely proportional to the square of the distance $r [r = (x^2 + y^2)^{0.5}]$. Establish the potential function, stream function, and streamline function.

8.7 A point source and a point sink locate at the left and right sides of the original point, respectively, and they are apart from the original point by 1m and their intensities are both $20 \text{m}^2/\text{s}$. Calculate (a) the velocity at the original point, (b) the value of ψ for the streamline that passes through point (0, 4) and the velocity at this point.

8.4 现有一流场，$v_x = kx^2$，$v_y = -2kxy$，其中 k 为常数。判定此速度场是否满足连续方程，此流场是否有旋？如果存在的话写出流函数、速度势函数和流线方程。

8.5 有一平面势流，其速度势为 $\varphi = K\theta$，其中 θ 是极角，K 是正常数。试计算：沿圆 $x^2 + y^2 = R^2$ 和 $(x-a)^2 + y^2 = R^2$ 的环量。

8.6 流体绕固定轴旋转，质点的角速度 ω 与离轴心距离 $r [r = (x^2 + y^2)^{0.5}]$ 的平方成反比，试确定其速度势、流函数及流线。

8.7 一点源和一点汇分别位于原点的左侧和右侧，均距离原点 1m。源和汇的强度均为 $20 \text{m}^2/\text{s}$。试计算：(a) 坐标原点处的速度，(b) 通过点 (0, 4) 的流线的 ψ 值及该点的速度。

Chapter 9　Flow Around Body Immersed

第9章　绕物流动

Flow around a solid body immersed in fluid is a very common phenomenon in engineering. We can see this kind of flows in many occasions, such as flow around a plate, the shell-side flow around the pipes in a heat exchanger, water flows around bridge pier, and so on. It is obvious that in these flow processes, solids will affect fluid flows, resulting in the velocity and pressure redistributions in flow field. Meanwhile, solid object will encounter the additional load arising from fluids. In this chapter, the boundary layer theorem will be introduced, as well as the flow velocity, pressure distribution and the load acting on object be analyzed.

工程上经常遇到的流体绕过物体的流动，如流体绕过平板、换热器壳程流体绕过换热管、河水绕过桥墩等的流动。显然，在绕流过程中，物体会影响流体的流动，导致流场中流体质点的速度和压力重新分布。同时流体也会对被绕流的物体产生作用力。本章主要介绍边界层理论，分析固体表面附近的流速、压力分布及流体对被绕流物体的作用力。

9.1　Overview of Boundary Layer

9.1　边界层概述

There will be a thin layer of fluid sticking to the surface for either a fluid flows around a solid body or a solid body moves in a fluid. In fluid mechanics, this thin layer is termed the boundary layer which is a layer of fluid in the immediate vicinity of a bounding surface.

无论是流体流过物体，还是物体在流体中运动，总有一薄层流体附着在物体表面上。流体力学中，这一紧贴壁面的薄层流体就定义为边界层。

Boundary layers can be loosely classified according to their structures and the circumstances under which they are created. The thin shear layer which develops on an oscillating body is an example of a Stokes boundary layer, while the Blasius boundary layer refers to the well-known similarity solution near an attached flat plate held in an oncoming unidirectional flow. When a fluid rotates and viscous forces are balanced due to the Coriolis effect, an Ekman boundary layer forms. In heat transfer, a thermal boundary layer occurs. Multiple types of boundary layers may coexist on a surface.

边界层可根据它们的结构以及形成因素进行大致分类。在振动物体表面形成的薄剪切层定义为 Stokes 边界层。Blasius 边界层指的是均匀来流掠过平板时贴近壁面的一层流体。当流体旋转且黏性力被 Coriolis 效应平衡时，会形成 Ekman 边界层。传热学当中还有传热边界层。固体表面上往往同时存在多种类型的边界层。

The velocity in the layer which sticks to boundary surface is in accordance to the surface velocity if there is no slip velocity, but there exist velocity differences among the external layers. As a result, viscous forces occur among the fluid layers.

As shown in Fig. 9.1, a horizontal plate moves at the velocity v_0 in a stationary fluid which is intentionally partitioned into thousands of thin layers. Due to the sticking effect of solid surface, the first layer will move with the wall together. The viscous force in the fluid will drive the second layer to move along the same direction as layer 1. But the inertia force will decelerate layer 2 so that there exists a velocity difference between v_1 and v_2. As for the other layers above layer 2, they are also driven to move horizontally, but their velocities decrease sequentially. Under steady state, the velocity distribution is like that shown in Fig. 9.1 and it is dependent on fluid viscosity. The velocity gradient in layer 1 is the largest comparing with that in other layers.

若无速度滑移，紧贴壁面的那层流体的速度与固体壁面一致。相邻的外层流体之间存在着速度差，从而使两层流体之间产生黏性力。

如图9.1所示，平壁在静止的流体中以速度 v_0 运动，将流体分为无数个薄层。由于壁面的黏附效应，第一层流体将以速度 v_0 随物体运动。在黏性力作用下，第2层流体将也沿着第1层流体的方向运动。但惯性力的作用使第2层流体速度减小，v_1 和 v_2 之间存在速度差。其他各层流体也将被拖动并沿水平方向运动，且速度依次降低。稳定后出现如图9.1所示的速度分布，这种分布与黏度有关。第一层内的速度梯度最大。

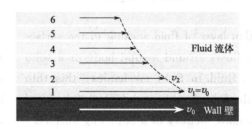

Fig. 9.1　Effect of viscosity on velocity distribution
图9.1　黏性对速度分布的影响

The viscous force in a fluid is related to both velocity gradient and its viscosity. In terms of the entire flow field, the viscous force effect is not obvious when the flow velocity is very large, and the energy loss due to viscous dissipation is relatively small. So the ratio of inertia force to viscous force (i.e. Re) just is the characteristic index of viscous fluid motion. As mentioned in chapter 6, when inertia force is large, Re is large and the effect of viscous force is weakened. Otherwise the effect of viscous force is remarkable.

黏性力与速度梯度和黏度有关。从整个流场来看，当流体的速度很大时，流体受黏性力的作用不大，由黏性而产生的能量损失也相对较小。所以惯性力与黏性力的比值（Re）才是全面描述黏性流体运动特征的指标。如第六章所述，惯性力大时，Re 大，黏性力的作用减小；反之，黏性力作用显著。

The evaluation of the effect of viscous force depends on several factors more than fluid viscosity. For a real fluid flow, such as air or water, the bulk flow can be regarded as the inviscid flow except for that in the very thin boundary layer. But if the velocity is low or the flow takes place in a passage with small characteristic length (i. e. the length in Reynolds number), viscous force can not be ignored.

Regardless of fluid viscosity and flow velocity, in the layer sticking to wall boundary, the velocity gradient is very large, so the role of viscous force is important in boundary layer. The velocity of the fluid in boundary layer is approximately parallel to the surface when the fluid flows along the surface. The fluid velocity is zero right on the boundary while it increases rapidly along the normal direction of the boundary, as can be seen in Fig. 9. 2, and the velocity gradient decreases.

仅凭流体的黏度大小，并不能决定其流动的黏性作用。例如，空气和水均是实际流体，除了在非常薄的边界层内，大部分可以看成是非黏性流动。但当流速很低或流道特征尺寸（Re 中的特征长度）很小时，则不能忽略黏性力。

不管流体黏度的大小、流速的高低，靠近物体表面处速度梯度很大，因而边界层中黏性力的作用很重要。流体沿固体表面流动时，边界层中的流体流速大致与物体表面平行。流速在物体表面为零，并沿外法线方向急剧增大，而速度梯度则减小，如图 9.2 所示。

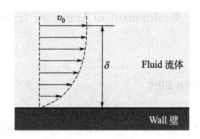

Fig. 9. 2 Velocity distribution near a fixed wall boundary
图 9.2 固定壁面边界附近的速度分布

As the Navier-Stokes equations describing the motion of viscous fluids can not be solved in most general cases, simplifying assumptions are used to describe approximate behavior. For a fluid with low viscosity, if the Reynolds number is large, the term about viscosity in the Navier-Stokes equation is a weakness compared to the inertial terms in majority of the flow filed. However, this is not true for near-wall fluid and this description fails to capture some important behavior in a lot of real systems, for instance the Kármán vortex street or the drag force on an airplane wing.

对于实际流体，直接采用纳维-斯托克斯方程式求解整个流场是很困难的，一般要进行一些假设。对于低黏度流体，当雷诺数较大时，黏性力较之于惯性力是小量，黏性力可以忽略。然而对于接近边界的区域，如果忽略黏性力则会丢失很多有用的信息。如后文会提到的卡门涡街现象和机翼上的阻力现象。

Through the analysis of the viscous force effect, Prandtl thought that a whole flow field can be discriminated between viscous flow field and inviscid flow field. The viscous region is referred to the boundary layer combined with its wake region. In this region the velocity field can be solved using the Navier-Stokes equations which can be simplified on the assumption that velocity vector is parallel to wall surface. The remainder is the inviscid region in which the viscous force is very small and the velocity gradient can be assumed to be zero. So it can be treated as a potential flow of ideal fluid.

9.2 Characteristics of boundary layer

9.2.1 Formation of Boundary Layer

The formation and development of a boundary layer are affected by both flow parameters and the shape of flow-passing surface. Some typical flows will be introduced to illustrate the development of boundary layer on wall surface.

(1) Flow in a shrinking pipe

Figure 9.3 illustrates a flow in shrinking pipe. The boundary layer is developing from position A to B and its thickness approaches the largest δ_C at point C, which corresponds to the inlet of the shrinking section. Once the fluid enters the shrinking section CD, its velocity increases while pressure reduces. Because the average velocity continually increases, the bulk flow provides more momentum to boundary layer and the velocity gradient in boundary layer increases, resulting in the reduce of boundary layer thickness to be δ_D. In the subsequent pipe section EF, the boundary layer develops again and the thickness increases correspondingly. Finally the boundary layer will develop to pipe center.

9.2 边界层特性

9.2.1 边界层的形成

流场中流动参量的变化、流道和绕流体形状的不同，都会影响边界层的形成和发展。下面介绍几种典型流动中的壁面边界层的发展。

（1）收缩管内的流动

图9.3所示为一收缩管中的流动。边界层从A处到B处不断发展，到C处后其厚度达到最大δ_C，而此处为收缩管的入口区。进入收缩管道CD段后，流体加速而压力逐渐降低。由于平均速度逐渐增高，主流对边界层流体的能量供应加强，使边界层内速度梯度增大，边界层逐渐减薄至δ_D。进入后续的直管段EF后，边界层又开始增厚，最后边界层会发展到管中心。

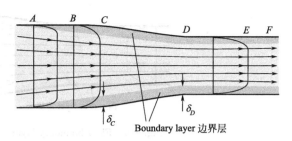

Fig. 9.3 Flow boundary layer in a shrinking pipe
图 9.3 收缩管内流动的边界层

(2) Flow around an aerofoil

The streamlines of the flow around an airfoil can be seen in Fig. 9.4. The boundary layer continually develops once the flow encounters the aerofoil and its thickness increases steadily, and a wake region forms at the back of the airfoil. At the nose of the wake region, the velocity gradient is large and finally the wake disappears in bulk flow.

（2）绕机翼流动

图 9.4 所示为均速流体绕过机翼的流动。边界层沿机翼表面发展并逐渐加厚，翼柱后部形成了尾迹区。尾迹区的鼻尖位置处速度梯度较大，随后尾迹逐渐扩散，最终消失在主流区中。

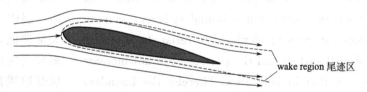

Fig. 9.4 Flow around an aerofoil
图 9.4 绕过机翼的流动

(3) Flow in a diverging pipe

As illustrated in Fig. 9.5, the boundary layer has developed for a distance before the fluid enters the diverging pipe. As the flow area increases, the pressure in bulk flow increases and the kinetic energy transforms into pressure energy. But the kinetic energy in boundary layer is not large enough to compensate the requirement of pressure energy. The kinetic energy arising from bulk flow is also reduced, and as a result, the velocity in boundary reduces to be zero and sometimes backflow occurs. Meanwhile bulk flow is still at a high velocity so that it separates from boundary layer, and vortex or backflow appears near wall boundary. In a word, the development of boundary layer in a diverging pipe may be unsteady and discontinuous.

（3）渐扩管中的流动

如图 9.5 所示，流体在进入渐扩管前，边界层已经历了一段距离的发展。随着流道截面的增大，主流区中动能转化为压力能。但边界层中的动能不足以补充压力能的增高，且主流对边界层流体能量供应减弱，致使边界层中流体的流速最终降为零，甚至出现倒流。此时主流仍以较高流速流动，主流与边界层分离，边界处出现旋涡和倒流等不规则的流动。因此，渐扩管中边界层的发展可能是不连续、不稳定的。

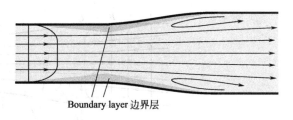

Fig. 9.5 Flow boundary layer in a diverging pipe

图 9.5 渐扩管内流动的边界层

(4) Flow around a cylinder

Fig. 9.6 shows a flow around a cylinder. The boundary layer goes on developing on the windward surface ($A \rightarrow B$ and $A \rightarrow C$) when fluid flows around the cylinder. In this course, the velocity increases while the pressure decreases because the flow area is decreasing, and the boundary layer will not separate from wall surface. In the leeward side of the cylinder ($B \rightarrow D$ and $C \rightarrow D$), velocity decreases while pressure increases because the flow area is increasing. But in boundary layer, most of the kinematic energy is depleted to resist the frictional force of wall surface and the pressure can not increase as large as that in bulk flow. Thereby the boundary layer separates from wall surface, and a series of vortexes form behind the cylinder and last for a distance which is several times of the cylinder diameter. A wake region can also be found in the flow field. This flow phenomenon will be introduced in detail in section 9.4.2.

(4) 绕圆柱流动

图 9.6 表示流体绕过圆柱体的流动。在圆柱体的迎流面上（$A \rightarrow B$ 和 $A \rightarrow C$），边界层逐渐形成并发展。此时由于流道面积减小，流体沿柱面是增速降压流动，不会出现边界层脱离现象。在圆柱体的背流区（$B \rightarrow D$ 和 $C \rightarrow D$），流动截面不断增大，流体作减速增压流动。边界层中因克服黏性摩擦而损失大量动能，无法补充足够的压力来与主流压力平衡。于是，边界层开始脱离固体表面，在圆柱背面形成一系列旋涡，并向下游发展，直到几倍圆柱直径的距离后消失。此流动中也能形成尾迹区。这一现象将在 9.4.2 节中详细介绍。

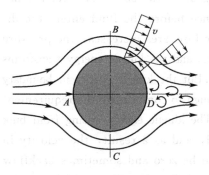

Fig. 9.6 Flow around a cylinder

图 9.6 绕圆柱流动

These four cases indicate that when a fluid flows over a wall surface, the boundary layer of viscous fluid may

可见，流体流过物体表面时，黏性流的边界层可能充分发展，也可

either fully develop or separate from boundary. Therefore, not all fluid apart from boundary layer can be regarded to fall in inviscid flow region. Only when Re is very large and no boundary layer separation occurs, the bulk flow can be treated as a potential flow, such as aerofoil moves in a steady field and fluid flows over a plane. It should be stated that the flow in a straight pipe is not an inviscid flow or potential flow because effect of viscous force extends from boundary to pipe center. That is different from the flow around a plate or curved surface.

9.2.2. The Laminar and Turbulent Boundary Layers

Laminar flow is dominated by viscous force and the velocity field is parallel to the boundary surface. But the velocity distribution in turbulent flow is time-dependent, so its velocity field usually refers to the distribution of time-averaged velocity. Rigorously the velocity in any flow filed always changes with time but if it varies in a small range and deviates a little apart from the time-averaged value, it can be considered as a laminar flow; otherwise it is a turbulent flow.

The transition from laminar flow into turbulent flow is dependent upon the occurrence and development of tiny turbulence. Tiny vortexes will be generated due to the velocity difference between two adjacent fluid layers with different velocities. These vortexes may be either decayed due to the decrease of velocity gradient initiated by viscous force or aggrandized due to the increase of velocity gradient initiated by inertia force. If vortexes vanish, the flow converts to be laminar; if vortexes grow continuously, the flow develops to be turbulent.

As shown in Fig. 9.7, once a uniform flow with the velocity of v_0 passes over a plate which is parallel to the flow direction, a thin boundary layer will form and grow from the leading end of the plate. Since the boundary layer is relatively thin near the leading edge, the turbulence can not develop easily in this region. So within this region the boundary layer appears as a smooth laminar flow and is dominated by viscous force. As flow goes on, the laminar boundary layer increases in thickness so that the turbulence develops and the thickness of the boundary layer increases quickly. Then the flow becomes turbulent and in this region the boundary layer is termed the turbulent boundary layer.

不易发展，所以此处属于层流边界层，受黏性力的控制。当流体沿平板继续流动，边界层逐渐增厚，扰动发展，边界层厚度迅速增加，边界层中的流动变成湍流，称为湍流边界层。

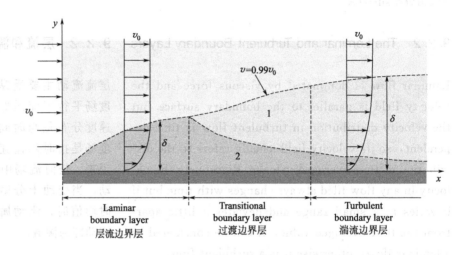

Fig. 9.7　Development of boundary layer on a plane

图 9.7　平板上边界层的发展

1—Turbulent sub-layer 湍流底层；2—Viscous sub-layer 黏性底层

At a position apart from the leading edge, the smooth laminar flow breaks down and transits to a turbulent flow. However in the very thin layer that sticking to wall boundary, the flow is still laminar and termed as the viscous sub-layer.

在远离平板前端的某一位置，光滑层流被破坏而转变为湍流。然而，与板面接触的薄层中仍为层流，称为黏性底层。

The transition from laminar flow to turbulent flow is dominated by the Reynolds number, Re. For a flow around a solid object, Re, is functional dependent on bulk flow velocity, v_0, dynamic viscosity μ and the distance apart from the leading edge, x_0, which holds

边界层由层流向湍流的转变，取决于雷诺数 Re。对绕流流场，Re 是主流速度 v_0、流体黏度 μ 和与板前端之间的距离 x_0 的函数，即

$$Re = \frac{\rho v x_0}{\mu} \qquad (9.1)$$

The critical value of Re_{cr} is related to the degree of initial turbulence in bulk flow, plate shape, pressure gradient in the flow, wall roughness, compressibility of fluid (the Mach number), so it is not a constant for different flows. For an adiabatic flow without pressure gradient on a smooth surface, the critical value Re_{cr} is

$$Re_{cr} = \frac{\rho v_0 x_{cr}}{\mu} = 3 \times 10^5$$

where x_{cr} is the critical flow distance.

If the turbulence arising from bulk fluid is very weak, Re_{cr} can approach 3×10^6. On this condition, the flow in laminar boundary layer is very smooth, while in turbulent boundary layer swirls or eddies do exist. The laminar boundary flow creates less frictional resistance than turbulent boundary flow, but is less stable.

If bulk flow velocity is constant, the occurrence of turbulent boundary layer can be avoided so long as the flow distance is smaller than critical distance, i.e., $x < x_{cr}$. x_{cr} should be determined before carrying out a boundary layer analysis.

The laminar boundary layer can be precisely analyzed according to the Newton's shear law $\left(\tau = \mu \dfrac{\mathrm{d}v_x}{\mathrm{d}y}\right)$. Even for nonplanar wall boundary, the analytical solution to velocity field can be obtained through a complex solving process. But for turbulent boundary layer, it is impossible to perform a quantitative calculation because there is no available physical model. In most engineering cases, the approximate solutions based on experimental results are usually employed for turbulent boundary flow. As for transitional boundary flow, it is much more complicated and is regarded as a superposition of laminar and turbulent flows or straightforwardly a pure turbulent flow.

主流的初始扰动程度、板面形状、流场的压力梯度、壁面的粗糙度、可压缩性（马赫数）等都会影响临界雷诺数 Re_{cr}。对于不同的流动，Re_{cr} 不是一个常数。对于光滑表面没有压力梯度的绝热流动

式中，x_{cr} 是临界的流动距离。

当主流扰动非常小时，Re_{cr} 可达到 3×10^6。层流边界层中的流动十分平缓，而湍流边界层中会存在旋转或旋涡。层流边界层流动的表面摩擦阻力要小于湍流，但并不稳定。

当主流速度一定时，只要流动距离小于临界值，即 $x < x_{cr}$，就可避免湍流边界层形成。在进行边界层分析前，首先要确定 x_{cr}。

对于层流边界层，根据黏性流的牛顿内摩擦定律 $\left(\tau = \mu \dfrac{\mathrm{d}v_x}{\mathrm{d}y}\right)$，可进行精确计算。对于非平面边界，仍能通过复杂的求解过程获得解析解。但对于湍流边界层，由于没有合适的物理模型，所以无法进行定量计算。对于大多数工程问题，一般采用基于实验结果的近似解。至于过渡边界层，流动更为复杂，在计算中往往近似地把它看成是层流和湍流的重叠区，或者全部按湍流来计算。

In the analysis of laminar or turbulent boundary layer, if $Re > 10^4$, the boundary layer thickness is a tiny comparing with the body size or the flow distance, x. On this condition, a simplified boundary layer computation can be performed. The bulk flow external to the boundary layer can be considered as an invicid flow.

9.3 Boundary Layer Equations

9.3.1 The Governing Equations

(1) Laminar boundary layer equations

Prandtl and Von Karman proposed a micron element analysis method and a control volume analysis method to establish the boundary differential equations and integral equations, respectively.

The deduction of the boundary layer equations was one of the most remarkable advances in fluid mechanics. Using the method of order-of-magnitude analysis, the Navier-Stokes equations were greatly simplified within the boundary layer. Notably, the characteristic of the partial differential equations (PDE) becomes parabolic, rather than the elliptical form of the full Navier-Stokes equations. This greatly simplifies the solution of the equations. By making the boundary layer approximation, the flow is divided into an inviscid portion (which is easy to solve by a number of methods) and the boundary layer, which is governed by an easier to solve PDE.

A boundary layer calculation can result in the boundary layer thickness, pressure distribution and flow drag force can be calculated.

The viscous fluid flow in boundary layer is governed by the continuity equation and the Navier-Stokes equations. By omitting the mass force, these equations for a two-dimensional steady and incompressible flow can be simplified as

$$\left.\begin{array}{l} v_x\dfrac{\partial v_x}{\partial x}+v_y\dfrac{\partial v_x}{\partial y}=-\dfrac{1}{\rho}\dfrac{\partial p}{\partial x}+\dfrac{\mu}{\rho}\left(\dfrac{\partial^2 v_x}{\partial x^2}+\dfrac{\partial^2 v_x}{\partial y^2}\right)\\ v_x\dfrac{\partial v_y}{\partial x}+v_y\dfrac{\partial v_y}{\partial y}=-\dfrac{1}{\rho}\dfrac{\partial p}{\partial y}+\dfrac{\mu}{\rho}\left(\dfrac{\partial^2 v_y}{\partial x^2}+\dfrac{\partial^2 v_y}{\partial y^2}\right)\\ \dfrac{\partial v_x}{\partial x}+\dfrac{\partial v_y}{\partial y}=0 \end{array}\right\} \quad (9.2)$$

Equation (9.2) can be further simplified according to the boundary layer characteristics. In the analysis of order-of-magnitude, the streamwise length is assumed to be significantly larger than the transverse thickness of boundary layer.

利用边界层特性对式(9.2)进一步进行简化在数量级分析中，假设流向长度要显著大于边界层的横向厚度。

As a result, the Navier-Stokes equations for a two-dimensional steady incompressible flow are obtained as

得到不可压缩黏性流体二维稳定流动的层流边界层方程

$$\left.\begin{array}{l} v_x\dfrac{\partial v_x}{\partial x}+v_y\dfrac{\partial v_x}{\partial y}=\dfrac{\mu}{\rho}\dfrac{\partial^2 v_x}{\partial y^2}\\ \dfrac{\partial p}{\partial y}=0\\ \dfrac{\partial v_x}{\partial x}+\dfrac{\partial v_y}{\partial y}=0 \end{array}\right\} \quad (9.3)$$

with the boundary conditions

其边界条件为

$$\left.\begin{array}{l} y=0, v_x=v_y=0\\ y=\delta, v_x=v_0(x) \end{array}\right\}$$

where $v_0(x)$ is the bulk velocity along flow direction x. When $\partial p/\partial x=0$, v_0 is independent of x.

式中，$v_0(x)$ 为沿 x 方向的主流速度。当 $\partial p/\partial x=0$ 时，主流流速与 x 无关。

The nonlinearity introduces much difficulty in the solution of these equations, even for those flows passing by the surface with simple geometric boundary.

由于边界层微分方程的非线性，即使对于简单形状的几何边界，求解也十分困难。

(2) Turbulent boundary layer equations

（2）湍流边界层方程

The investigation on turbulent boundary layers is far more difficult due to the time-dependent variation of the flow properties. Usually the instantaneous flow velocity

由于流动特性参数的时间相关性，湍流边界层的处理比层流要复杂得多。一般采用时均值加波动量

is decomposed into a time-averaged value and a fluctuation value. Thus the full turbulent boundary layer equations becomes

来代替瞬时值，因此，得到的边界层方程变为

$$\left.\begin{array}{l}\overline{v}_x\dfrac{\partial \overline{v}_x}{\partial x}+\overline{v}_y\dfrac{\partial \overline{v}_x}{\partial y}=-\dfrac{1}{\rho}\dfrac{\partial \overline{p}}{\partial x}+\dfrac{\mu}{\rho}\left(\dfrac{\partial^2 \overline{v}_x}{\partial x^2}+\dfrac{\partial^2 \overline{v}_x}{\partial y^2}\right)-\dfrac{\partial}{\partial y}(\overline{v'_x v'_y})-\dfrac{\partial}{\partial x}(\overline{v'^2_x})\\ \overline{v}_x\dfrac{\partial \overline{v}_y}{\partial x}+\overline{v}_y\dfrac{\partial \overline{v}_y}{\partial y}=-\dfrac{1}{\rho}\dfrac{\partial \overline{p}}{\partial y}+\dfrac{\mu}{\rho}\left(\dfrac{\partial^2 \overline{v}_y}{\partial x^2}+\dfrac{\partial^2 \overline{v}_y}{\partial y^2}\right)-\dfrac{\partial}{\partial x}(\overline{v'_x v'_y})-\dfrac{\partial}{\partial y}(\overline{v'^2_y})\\ \dfrac{\partial \overline{v}_x}{\partial x}+\dfrac{\partial \overline{v}_y}{\partial y}=0\end{array}\right\} \quad (9.4)$$

where the variables with symbol "$-$" present time-averaged values and those with symbol "$'$" present fluctuation values.

式中，带"$-$"符号的变量表示时均值；带"$'$"符号的表示波动值。

Also using an order-of-magnitude analysis, the above equations can be reduced to leading order terms. The momentum equation in x direction simplifies to

同样，通过采用数量级分析，保留大数量级参量，可将 x 方向的动量方程简化为

$$\overline{v}_x\dfrac{\partial \overline{v}_x}{\partial x}+\overline{v}_y\dfrac{\partial \overline{v}_x}{\partial y}=-\dfrac{1}{\rho}\dfrac{\partial \overline{p}}{\partial x}-\dfrac{\partial}{\partial y}(\overline{v'_x v'_y})$$

This equation does not satisfy the no-slip velocity condition. Like Prandtl did for his boundary layer equations, a new, smaller length scale was used to allow the viscous term to become leading order in the momentum equation. By choosing $\eta \ll \delta$ as the y-scale, the leading order momentum equation for this "inner boundary layer" is given by

但这一方程不满足无滑移边界条件。Prandtl 提出采用一个更小的尺度标准，从而将黏性项的尺度作为动量方程的主导尺度。采用 η ($\eta \ll \delta$) 作为 y 方向尺度，内边界层方程的动量方程变为

$$0=-\dfrac{1}{\rho}\dfrac{\partial \overline{p}}{\partial x}+\nu \dfrac{\partial^2 \overline{v}_x}{\partial y^2}-\dfrac{\partial}{\partial y}(\overline{v'_x v'_y})$$

The new "inner length scale", η, is a viscous length scale, and is of order ν/v^*, with v^* being the velocity scale of the turbulent fluctuations, in this case a friction velocity.

新的尺度 η 称为黏性长度比尺，与 ν/v^* 同一数量级，其中 v^* 表示湍动程度，此处为摩擦速度。

The additional term $\overline{v'_x v'_y}$ is known as the Reynolds shear stress. The solution of the turbulent boundary layer equations therefore necessitates a turbulence model, which aims to express the Reynolds shear stress in terms of known flow variables or derivatives.

附加项 $\overline{v'_x v'_y}$ 即是雷诺切应力。求解湍流边界层方程需要用到湍流模型，然后用已知的流动参数来表示雷诺切应力。

9.3.2 Boundary Layer Thickness

For a turbulent pipe flow or a flow around solid body, large velocity gradient occurs in the thin layer sticking to wall boundary. At wall surface the fluid satisfies a no-slip boundary condition thus the boundary velocity is zero. Along the normal direction the velocity increases continuously and approaches the velocity of bulk flow. Since the velocity variation is continuous, it is impossible to define an interface at which the boundary layer connects to bulk flow. The parameters below provide a useful definition of this characteristic, measurable thickness.

(1) 99% boundary layer thickness

The overall boundary layer thickness (or height), δ, is the distance across a boundary layer from the wall to a point where the flow velocity has essentially reached the free stream velocity, v_0. This distance is defined normal to the wall, and the point is customarily defined as the point where

$$v(y) = 0.99 v_0 \quad (9.5)$$

For the laminar boundary layer over a flat plate, the Blasius solution gives

$$\delta \approx 4.91 \sqrt{\frac{\nu x}{v_0}} \quad \text{or} \quad \delta \approx 4.91 x / \sqrt{Re_x}$$

For turbulent boundary layers over a flat plate, the boundary layer thickness is given by

$$\delta \approx 0.382 x / Re_x^{1/5}$$

where $Re_x = \dfrac{\rho v_0 x}{\mu}$.

(2) Displacement thickness

The displacement thickness, δ_1, is the distance by

9.3.2 边界层厚度

在管内湍流和绕流情况下，流场中的速度变化主要发生在壁面附近。无速度滑移时，紧贴壁面的流体速度为0，且沿法线方向速度逐渐增大，最终等于主流速度。由于速度是连续变化的，因此不存在速度转折点，因而就无法定义边界层与主流的界面。下面列出了几种规定边界层厚度的方法。

(1) 99%速度边界层

自壁面至流速基本达到主流速 v_0 处的距离称为速度边界层厚度（或高度），用 δ 表示。这段距离是沿着壁面法向的，一般将速度上限点定义为

对于层流边界层，其厚度的 Blasius 解为

对于平板湍流边界层，其厚度为

式中，$Re_x = \dfrac{\rho v_0 x}{\mu}$。

(2) 位移厚度

位移厚度 δ_1 的定义是：对于速度

which a surface would have to be moved in the direction perpendicular to its normal vector away from the reference plane (at x-axis) in an inviscid fluid stream of velocity v_0, so as to give a reduce of flow rate which is the same as that initiated by decrease of velocity due to the sticking effect in a real fluid flow. As shown in Fig. 9.8, the total area of region 1 and region 3 equals to that of region 2 and region 3.

为 v_0 的无黏（理想）流体流动，将流动边界从参考面（位于 x 轴上）平移 δ_1 的距离后，减少的流量应等于实际流体由于黏滞效应引起速度减小而使整个流场减小的流量。如图 9.8 所示，即面积 1＋面积 3＝面积 2＋面积 3。

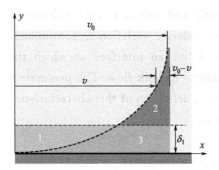

Fig. 9.8　Diagram of displacement thickness
图 9.8　位移厚度示意图

The definition of the displacement thickness for a compressible flow is based on mass flow rate, i.e.

可压缩流体的位移厚度定义为

$$\delta_1 = \int_0^\infty \left[1 - \frac{\rho(y)v(y)}{\rho_0 v_0}\right] dy \quad (9.6a)$$

The definition for incompressible flow can be based on volumetric flow rate, since the density is constant

不可压缩流体的位移厚度定义为

$$\delta_1 = \int_0^\infty \left[1 - \frac{v(y)}{v_0}\right] dy \quad (9.6b)$$

where ρ_0 and v_0 are the density and velocity in the free stream outside the boundary layer, and y is the coordinate normal to the wall.

式中，ρ_0 和 v_0 分别是边界层外主流的密度和速度；y 是壁面的法向坐标。

(3) Momentum thickness

（3）动量厚度

The momentum thickness, δ_2, is the distance by which a surface would have to be moved parallel to itself away from the reference plane in an inviscid fluid stream of velocity v_0, so as to give a reduce of momentum which is the same as that initiated by decrease of velocity due to the sticking effect in a real fluid flow.

动量厚度 δ_2 的定义是：对于速度为 v_0 的理想流体流动，将流动边界从参考面平移 δ_2 的距离后，减少的动量应等于实际流体由于黏滞效应引起速度减小而使整个流场减小的动量。

The definition of the momentum thickness for compressible flow is based on mass flow rate

可压缩流体的动量厚度定义为

$$\delta_2 = \int_0^\infty \frac{\rho(y)v(y)}{\rho_0 v_0}\left[1 - \frac{v(y)}{v_0}\right]\mathrm{d}y \quad (9.7\mathrm{a})$$

The definition for incompressible flow can be based on volumetric flow rate, as the density is constant

不可压缩流体的动量厚度定义为

$$\delta_2 = \int_0^\infty \frac{v(y)}{v_0}\left[1 - \frac{v(y)}{v_0}\right]\mathrm{d}y \quad (9.7\mathrm{b})$$

For a flat plate with zero angle of attack with a laminar boundary layer, the Blasius solution gives

对于攻角为零的平板流动，层流边界层的 Blasius 解为

$$\delta_2 \approx 0.664\sqrt{\frac{\nu x}{v_0}} \quad (9.7\mathrm{c})$$

(4) Energy thickness

（4）能量厚度

The energy thickness, δ_3, is the distance by which a surface would have to be moved parallel to itself towards the reference plane in an inviscid fluid stream of velocity v_0, so as to give a reduce of kinetic energy which is the same as that initiated by decrease of velocity due to the sticking effect in a real fluid flow. It is expressed as

能量厚度 δ_3 的定义是：对于速度为 v_0 的理想流体流动，将流动边界从参考面平移 δ_3 的距离后，减少的动能应等于实际流体由于黏滞效应引起速度减小而使整个流场减小的动能。可表示为

$$\rho\frac{v_0^3}{2}\delta_3 = \int_0^\infty \rho v\left(\frac{v_0^2}{2} - \frac{v^2}{2}\right)\mathrm{d}y$$

Thereby

因此

$$\delta_3 = \int_0^\infty \frac{v}{v_0}\left(1 - \frac{v^2}{v_0^2}\right)\mathrm{d}y \quad (9.8)$$

(5) Shape factor

（5）形状系数

A shape factor, H, is used in boundary layer flow to determine the nature of the flow, which holds

形状系数 H 用来表征边界层流动的本质

$$H = \frac{\delta_1}{\delta_2}$$

The higher the value of H, the stronger the adverse pressure gradient. A high adverse pressure gradient

H 越大，边界层中的反压力梯度越大。反压力梯度越大，由层流至

can greatly reduce the critical Reynolds number at which transition into turbulence may occur. Conventionally, $H = 2.59$ (Blasius boundary layer) is typical of laminar flows, while $H = 1.3 \sim 1.4$ is typical of turbulent flows.

9.4 Flow Around a Cylinder

9.4.1 Ideal Fluid Flow Around a Cylinder

An infinitely long cylinder with the radius of r_0 is transversely immersed in a uniform flow of ideal fluid. This is a typical planar flow of ideal fluid around a cylinder, as shown in Fig. 9.9. The streamlines are uniform when they are far away from the cylinder, and in the region near the cylinder they are winding.

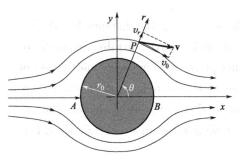

Fig. 9.9 Ideal fluid flow around a cylinder
图 9.9 理想流体绕圆柱流动

The fluid particles which move along the x axis will collide with the cylinder at the front stag point A where the velocity is zero. Then the fluid will be divided into two streams which move along the cylinder profile up and down, respectively. At point B, which is the back stag point, the two streams will collide with each other and both of their velocities become zero. Then the particles in the two streams will move along the x direction with free streams.

As introduced in chapter 8, the ideal fluid flow around a cylinder is a superposition of a uniform linear flow and a doublet flow with the original point of the coordinate system being the doublet center.

The potential function and stream function of uniform linear flow along the x axis with the velocity of v_∞ are

速度为 v_∞、与 x 轴平行的均匀直线流动的势函数与流函数分别为

$$\left.\begin{array}{l}\varphi_1=v_\infty x \\ \psi_1=v_\infty y\end{array}\right\} \quad (9.9)$$

and the potential function and stream function of doublet flow are

偶极流的势函数与流函数分别为

$$\left.\begin{array}{l}\varphi_2=\dfrac{Mx}{2\pi(x^2+y^2)}=\dfrac{Mx}{2\pi r^2} \\ \psi_2=-\dfrac{My}{2\pi(x^2+y^2)}=\dfrac{-My}{2\pi r^2}\end{array}\right\} \quad (9.10)$$

Superposition of the potential functions and stream functions gives

叠加后的复合流动的势函数和流函数为

$$\left.\begin{array}{l}\varphi=v_\infty x+\dfrac{Mx}{2\pi r^2} \\ \psi=v_\infty y-\dfrac{My}{2\pi r^2}\end{array}\right\} \quad (9.11)$$

$\psi=0$ presents the zero streamline and two solutions of Eq. (9.11) can be obtained, i.e., $y=0$ and $r=\sqrt{\dfrac{M}{2\pi v_\infty}}$. The former denotes the x axis and the later denotes the cylindrical surface. When $r=r_0$ the dipole moment is

流函数 $\psi=0$ 表示一条零流线，由式(9.11)可得到两个解：$y=0$ 及 $r=\sqrt{\dfrac{M}{2\pi v_\infty}}$，前者表示 x 轴，后者为一圆柱面。当 $r=r_0$ 时偶极矩为

$$M=2\pi v_\infty r_0^2 \quad (9.12)$$

Substituting Eq. (9.12) into Eq. (9.11) yields

将式(9.12)代入式(9.11)得

$$\varphi=v_\infty x\left(1+\dfrac{r_0^2}{r^2}\right)=v_\infty \cos\theta\left(r+\dfrac{r_0^2}{r}\right) \quad (9.13)$$

$$\psi=v_\infty y\left(1-\dfrac{r_0^2}{r^2}\right)=v_\infty \sin\theta\left(r-\dfrac{r_0^2}{r}\right) \quad (9.14)$$

The velocity components are

其速度分量为

$$v_x=\dfrac{\partial \varphi}{\partial x}=v_\infty-v_\infty r_0^2\dfrac{x^2-y^2}{(x^2+y^2)^2}$$

$$v_y=\dfrac{\partial \varphi}{\partial y}=-v_\infty r_0^2\dfrac{2xy}{(x^2+y^2)^2}$$

When $r \to \infty$, $v_x \to v_\infty$ and $v_y \to 0$, which indicates that when it is far away from the cylinder, the velocity, v_x, is parallel to the x axis and equals to v_∞.

The radial and tangential velocity components can also be obtained from Eq. (9.13), such as

$$v_r = \frac{\partial \varphi}{\partial r} = v_\infty \left(1 - \frac{r_0^2}{r^2}\right) \cos\theta$$

$$v_\theta = \frac{1}{r}\frac{\partial \varphi}{\partial \theta} = -v_\infty \left(1 + \frac{r_0^2}{r^2}\right) \sin\theta$$

The above two equations show that when $r = r_0$, $v_r = 0$ and $v_\theta = -2v_\infty \sin\theta$. It can be understood that velocity on the cylinder surface is always along the tangential direction of cylinder surface, which means the flow will not separate from the surface. On the front stag point A ($\theta = \pi$), $v_\theta = 0$, on the back stag point B ($\theta = 0$), $v_\theta = 0$, and when $\theta = \pm\pi/2$, $v_\theta = \mp 2v_\infty$.

The pressure distribution along the cylindrical surface can be determined using the Bernoulli equation. Supposing the static pressure at an infinite point is p_∞, the velocity and pressure on the cylindrical surface are v_S and p_S respectively, Eq. (9.15) can be obtained by neglecting the potential energy difference

$$p_\infty + \frac{\rho}{2}v_\infty^2 = p_S + \frac{\rho}{2}v_S^2 \quad (9.15)$$

i.e.

$$p_S = p_\infty + \frac{\rho}{2}v_\infty^2 (1 - 4\sin^2\theta) \quad (9.16)$$

In engineering a dimensionless pressure $\bar{p}_S$ is usually employed, which is defined as

$$\bar{p}_S = \frac{p_S - p_\infty}{\frac{\rho}{2}v_\infty^2}$$

Thereby the dimensionless pressure distribution along the cylindrical surface can be expressed as

$$\bar{p}_S = 1 - 4\sin^2\theta \qquad (9.17)$$

It can be seen that $\bar{p}_S$ is not affected by v_∞ or p_∞.

可见，$\bar{p}_S$ 不受 v_∞ 及 p_∞ 的影响。

9.4.2 Viscous Fluid Flow Around a Cylinder

(1) Flow separation

Due to viscosity effect of real fluid, boundary layer will form on cylinder surface. The impediment of the cylinder will slow down the fluid and the velocity at the front stag point is zero. Here the boundary layer does not develop so its thickness is small. On the windward side of the cylinder, the velocity is increasing that hinders the increase of boundary layer thickness. The flow in boundary layer is still along the direction of free stream which provides the momentum to make up the energy consumption resulted from the viscous dissipation in boundary layer. Hence the velocity in boundary layer will not reduce and the flow in boundary layer remains steady although it grows continuously.

Once the fluid climbs over the highest point of cylinder (point A in Fig. 9.10), the velocity decreases while the pressure increases due to the increase of flow area. There exists a reverse pressure gradient with respect to the flow direction, and the free stream becomes farther away from the leeward surface of the cylinder. The kinetic energy of free stream mainly transforms to pressure

9.4.2 黏性流体绕圆柱流动

(1) 流动分离

由于实际流体存在黏性，在柱体表面上形成边界层。当流体流近柱体前驻点时，流体速度逐步降低，至前驻点处流体速度为零，该处边界层尚未发展，边界层厚度较薄。当流体绕柱体前半圆周时，沿流向速度越来越大，减缓了边界层厚度的增加。由于边界层内由流体黏性摩擦引起的能量消耗可由外侧主流补充，使边界层内的速度不会降低，因此边界层虽然不断增长，却是稳定的。

但当流体绕过柱体最高点 A 后 (图 9.10)，由于通流面变大，主流变成增压减速流动。此时出现与主流方向相反的逆压力梯度，且主流远离固体表面。这时主流的动能要转化为压力能，速度不断下降，无力补充边界层中能量

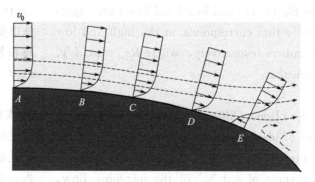

Fig. 9.10 Boundary layer development on the back of cylinder

图 9.10 圆柱体背侧边界层的发展

energy and is unable to compensate the energy dissipation in boundary layer. Meanwhile the boundary layer thickness increases larger and larger.

As the flow area increases rapidly, the pressure increases correspondingly and at a point on the leeward side (point B on the surface), some fluid in boundary layer stop and the succeeding fluid piles up around the point. The free stream has to move around the pileup and separates from the wall boundary, as shown in Fig. 9.11(a). Thereby the boundary layer separation takes place. As the free stream separate from the boundary, the piled up fluid is injected into the free stream and eddies form behind the cylinder, as Fig. 9.11(b) illustrates, then point B is not the stag point anymore. On this condition, the pressure distribution along the surface can not be expressed by Eq. (9.17).

的消耗，于是边界层的厚度却越来越大。

随着通流面的快速增大，压力继续增大，于是背流面某处（B 点）的流体完全静止，而后续流体在此处堆积。主流必须绕过这部分堆积物，这时便发生了边界层脱离，如图 9.11(a) 所示。主流脱体后，向圆柱体两侧分流，并引射边界层堆积的流体物质进入主流，使柱体背后形成涡流区，如图 9.11(b) 所示。这时 B 点不再是驻点，沿圆柱表面压力分布也不再符合式(9.17)的规律。

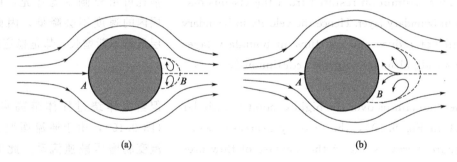

Fig. 9.11 Flow separation
图 9.11 流动分离

Figure 9.12 shows the dimensionless pressure distribution along the cylindrical surface. Line 1 is drawn according to Eq. (9.16) and line 2 and line 3 are experimental results that corresponds to the high and low Reynolds numbers respectively, where $Re_d = v_\infty d/\nu$. It can be seen that:

① There exist obvious differences between the pressure distributions of ideal fluid and viscous fluid along the cylindrical surface. But it can be found that within the illuminating angle of $\alpha \pm 30°$ of the incoming flow, their difference is small.

图 9.12 所示为沿圆柱表面的无因次压力分布线。曲线 1 根据式 (9.16) 计算得到；曲线 2、曲线 3 是由实验测得的，分别对应于较高和较低的绕流雷诺数（$Re_d = v_\infty d/\nu$）。可见：

① 实际流体绕流与理想流体绕流的柱面压力分布曲线有一定差别。只是在与来流方向成 30°的区域内，即 $\alpha \pm 30°$时，两者的区别很小。

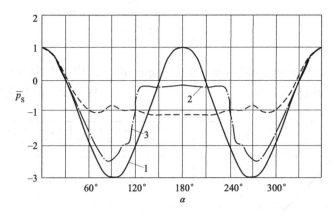

Fig. 9.12 Dimensionless pressure distribution along the cylindrical surface
图 9.12 沿圆柱体表面上的无量纲压力分布

1—ideal fluid 理想流体；2—Re_d larger than critical value Re_d 大于临界值（$Re_d = 6.7 \times 10^5$）；
3—Flow with low Re_d Re_d 较低的流动（$Re_d = 1.86 \times 10^5$）

② For an ideal fluid flow around a cylinder, the resultant force acting on the cylinder surface is zero regardless of the flow speed. But for viscous fluid flow, there exists a pushing force on the cylinder along the flow direction, which is induced by the pressure difference between the windward side and leeward side.

③ Pressure distribution on the cylinder surface is related to the Reynolds number Re_d and boundary layer characteristics. For a low Reynolds number, the boundary layer behaves a laminar flow, the separation occurs more early and the pressure curve is more complanate. If the Reynolds number exceeds the critical value, the flow in boundary layer becomes turbulent. Since the momentum exchange between the free stream and the turbulent boundary layer is stronger, the fluid in boundary layer gets more energy to resist the sticking effect of wall boundary and then the separation point moves back. As a result, the flow state is improved compared with that under low Re_d.

It can be seen from Fig. 9.13 that when $Re_d > 2 \times 10^5$, the drag coefficient of viscous fluid flow around cylinder, C_f, reduces rapidly. C_f is defined as

② 对理想流体绕流，不管流体速度多大，整个圆柱体在流体中所受合力为 0。对实际流体绕流，流体对被绕流的柱体有一个沿流向的推力，由柱体前侧指向后侧，这是由柱体前后压差引起的。

③ 压力分布与绕流的雷诺数 Re_d 以及柱体表面上边界层的性质有关。当绕流雷诺数较低时，柱体表面的边界层属于层流边界层，脱离点较靠前，压力分布线较平坦。当 Re_d 超过临界值时，边界层由层流转变为湍流。由于湍流边界层与主流进行动量交换的能力更强，边界层获得了更多能量以克服黏性阻滞，脱离点向后推移，与低 Re_d 情况相比流动状态得以改善。

由图 9.13 可见，当 $Re_d > 2 \times 10^5$ 时，黏性流体绕流圆柱体的阻力系数 C_f 迅速下降。阻力系数定义为

$$C_f = \frac{D}{A \frac{\rho}{2} v_\infty} \quad (9.18)$$

where A is the maximum upwind area (A = length of cylinder × diameter of cylinder).

式中，A 为最大迎流面积，A = 圆柱长度×圆柱直径。

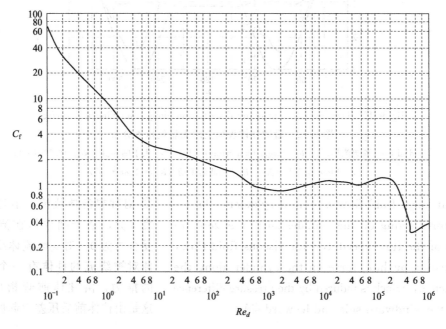

Fig. 9.13　Drag coefficient of viscous fluid flow around cylinder

图 9.13　黏性流体绕圆柱流动的阻力系数

Boundary layer separation can also be found in pipe flows. It can result from such a case as a rapidly expanding duct of pipe. Separation occurs due to an adverse pressure gradient encountered as the flow expands, causing an extended region of separated flow.

管内流动同样会出现边界层分离。如突然扩大的管道中，管道的扩大引起了逆压力梯度，继而产生流动分离，造成分离区域扩大。

Boundary layer control refers to methods of controlling the behavior of boundary layer, which is of great interest in aeronautical engineering because aerodynamic drag may be reduced whilst achieving high lift.

边界层控制指的是控制边界层内的流动特性，此类方法在航空领域受到极大重视，可以减小空气动力学阻力并获得大的升力。

For a free stream flows around a cylinder, several methods may be employed to control the boundary layer separation that occurs due to the adverse pressure gradient. First is to rotate the cylinder to reduce or eliminate the boundary layer; Second is to generate a reverse

对于圆柱绕流，有多种方法可以控制由逆压力梯度引起的边界层分离。一是使圆柱体旋转以削减或消除边界层；二是使侧边产生一个相对于主流的逆向边界层流动，

boundary layer flow with respect to the free stream, resulting in a partial separation of a boundary layer; third is the suction applied through a slit in the cylinder near the separation point, which can also delay the onset of separation by removing fluid particles that have been slowed in the boundary layer. Alternatively, fluid can be blown from a faired slit such that the slowed fluid is accelerated and thus the point of separation is delayed.

这样只有部分边界层会发生分离。也可在圆柱体分离点处开一个沟槽以抽吸该处的流体，通过去除低速的流体而将分离点推后。与抽吸相仿的是采用吹扫的方法提高低速流体的速度，这样也可以使分离点推后。

(2) Kármán vortex street

(2) 卡门涡街

When boundary layer separation occurs, the displacement thickness increases quickly, that modifies the outside potential flow and pressure field. For example, the boundary layer separation on an airfoil will induce pressure field change and increase the pressure drag, and if severe enough will also result in loss of lift and stall, all of which are undesirable. For internal flows, flow separation produces an increase in the flow losses, and stall-type phenomena such as compressor surge, both undesirable phenomena.

边界层发生分离后，位移厚度迅速增加，这将改变外部势流和压力场。如果机翼上发生边界层分离，压力场的改变将增加压差阻力，若情况进一步恶化，会导致失去升力或失速等恶劣后果。对于内流，边界层分离会导致流阻增加、失速型现象（如压缩机的浪涌）等不良现象的出现。

Another effect of boundary layer separation is shedding vortices, known as the Kármán vortex street. In fluid dynamics, a Kármán vortex street is a repeating pattern of swirling vortices caused by the unsteady separation of flow of a fluid around blunt bodies.

边界层分离的另一个影响是发生旋涡脱落，即生成卡门涡街。卡门涡街实际上就是由于流动的不稳定分离，在钝绕流物后方交替出现的一系列旋涡。

Fig. 9.14 Kármán vortex street
图 9.14 卡门涡街

Figure 9.14 shows the Kármán vortex street observed in experiment, and a sequence of vortexes are released regularly at the back of the cylinder. A vortex street

图 9.14 所示为实验观察得到的卡门涡街，一系列旋涡在圆柱背后有规则的脱落。卡门涡街只在一定

will only form at a certain range of flow velocities, specified by a range of Reynolds numbers (Re), typically above a limiting Re_d value of about 90.

The vortex street was investigated by von Kármán in 1911 and he proposed the conditions for generating a steady vortex street, as illustrated in Fig. 9.15. It should be a two-dimensional ideal fluid flow around an infinitely long cylinder and the rotation direction of vortexes at one side should be reversed to that at the other side. Furthermore, the flow external to the vortexes should be potential flow and infinite. On these conditions, the vortexes, as shown in Fig. 9.15, are steady. The separation distance of vortexes should meet the requirement of $h/l > 0.281$.

的速度范围内出现,可用雷诺数来判断。一般来说当 $Re_d > 90$,开始形成涡街。

冯·卡门于1911年研究了这种涡街现象并提出了涡街产生的条件,如图9.15所示。假设这是一个理想流体绕无限长圆柱体流动(平面问题),且两侧旋涡的旋转方向是相反的,旋涡以外的流动都是势流,而且是无限的。由这些假定推得,只有当旋涡按图9.15所示的方式排列时才是稳定的。稳定释放的涡街间距要求为 $h/l > 0.281$。

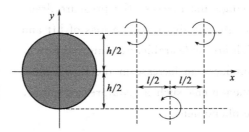

Fig. 9.15 Conditions for a steady Kármán vortex street
图9.15 稳定卡门涡街的条件

In the range of Re_d from 200 to 50000, the shedding off frequency f_K is proportional to v_∞ while inversely proportional to the diameter of cylinder, d.

在 $Re_d = 200 \sim 50000$ 范围内,柱后涡流的脱落频率 f_K 和 v_∞ 成正比,而和 d 成反比,可用下式表示

$$f_K = St \frac{v_\infty}{d} \qquad (9.19)$$

When $Re_d = 300 \sim 2 \times 10^5$, $St = 0.20$. As Re_d increases, the shedding point moves back and the width of the vortex street reduces that results in the increase of St. When Re_d is larger than 3.3×10^5, the wake flow becomes disorder and regular shedding off of vortexes disappear. But once $Re_d > 3.5 \times 10^6$, the vortex street is reproduced ($St = 0.27$). Being related to the Reynolds number, the shedding off frequency, f_K, is in the form

当 $Re_d = 300 \sim 2 \times 10^5$ 时,$St = 0.20$。随着 Re_d 增大,脱离点后移,涡街宽度减小,St 增加。当 $Re_d > 3.3 \times 10^5$ 时,尾流完全紊乱,有规则的涡流脱落消失。一旦 $Re_d > 3.5 \times 10^6$,又会形成卡门涡流,这时 $St = 0.27$。脱落频率 f_K 与雷诺数相关,为

$$f_K = 0.20 \frac{v_\infty}{d}\left(1 - \frac{20}{Re_d}\right) \quad (9.20)$$

Accompanied with the shedding off of vortexes, a periodic alternating crosswind force [F_K in Eq. (9.21)] acts on the cylinder and its frequency is the same as the shedding frequency which can approach hundreds of hertz.

$$F_K = C_K \frac{\rho}{2} v_\infty^2 A \sin\omega t \quad (9.21)$$

where C_K is the Kármán force coefficient (when $Re = 10^2 \sim 10^7$, $C_K = 1.0$), A is the projection area of cylinder, ω is circular frequency and $\omega = 2\pi f_K$.

If the shedding off frequency is close to the natural frequency of the cylinder, the resonance of the cylinder will be activated and the vibration amplitude of the cylinder will become larger and larger until it is destroyed.

For a periscope equipped on a submarine, if the frequency of the transverse excited force is equal to the natural frequency of the periscope by chance, the periscope will be damaged. The voice one hears from a power line blown by wind is also resulted from the vibration of the line induced by the periodic transverse force.

Therefore periodic crosswind forces set up by vortices along object's sides can be highly undesirable, and hence it is important for engineers to account for the possible effects of vortex shedding.

In order to prevent the undesired vibration of such cylindrical bodies, a longitudinal fin can be fitted on the downstream side, which, providing it is longer than the diameter of the cylinder, will prevent the eddies from interacting. Obviously, for a tall building or mast, the relative wind could come from any direction. For this reason, helical projections are sometimes placed at the top, which effectively create asymmetric three-

涡流脱落时，流体对圆柱体有一个周期性交变的作用力 F_K，如式 (9.21)，其频率与涡流脱落频率相同，频率可达几百赫兹。

式中，C_K 是卡门力系数（当 $Re = 10^2 \sim 10^7$ 时，$C_K = 1.0$）；A 是圆柱的投影面积；ω 是圆频率，$\omega = 2\pi f_K$。

如果横向交变力的频率与圆柱体的固有频率相等，就会引起柱体的共振，使柱体的振幅越来越大，直到破坏。

对于潜艇上伸出水面的潜望镜，如果镜筒受到的横向激动力的频率与其固有频率一致，则将造成潜望镜共振破坏。风吹电线嘘嘘发响的现象也是由卡门涡流周期性脱落时引起风压脉动所造成的。

因此，由旋涡脱落产生的周期性横向力应尽力避免。工程师在设计时都应考虑到旋涡脱落产生的影响。

为了避免圆柱体产生不必要的振动，可在背面加装经向的翅片，从而使圆柱体背流面直径变大，以阻止涡列的相互干扰。对于高建筑物来说，风可能来自四面八方，于是可在顶部加装螺旋式凸出物，这样可以构造有效的三维流动，以避免旋涡的交互脱落。另

dimensional flow, thereby discouraging the alternate shedding of vortices; another countermeasure with tall buildings is using variation in the diameter with height, such as tapering—that prevents the entire building being driven at the same frequency.

9.5 Flow Around a Sphere

The flow around a sphere is a spatial problem. There is no appropriate stream function for spatial flow so that investigation on spatial flow is much more difficult than planar flow. Considering an axisymmetric flow can be described by a stream function, it is advantageous to use the cylindrical coordinate system (r, φ, z) in which the origin is set at the sphere center, the z-axis passes through the sphere center and aligns with bulk flow direction and r is the radius as measured perpendicular to the z-axis. Because the flow is axisymmetric about the z-axis, it is independent of φ. Thus we can approximately analyze this problem by taken this flow as a planar potential flow.

9.5.1 Ideal Fluid Flow Around a Sphere

The analysis of a spatial flow around a sphere is related to spatial point source, point sink and doublet flows. For a spatial point source flow, the intensity of Q on the spherical surface with the radius of R can be calculated by

$$Q = 4\pi R^2 v_r = 4\pi R^2 \frac{\partial \varphi}{\partial R}$$

The velocity potential of a spatial point source flow is

$$\varphi = -\frac{Q}{4\pi R} \quad (9.22)$$

Similarly, for a point sink flow, the velocity potential is

$$\varphi = \frac{Q}{4\pi R} \quad (9.23)$$

9.5 绕球流动

流体绕圆球流动是一个空间问题。由于空间流动找不到合适的流函数，因此空间问题的研究要比平面问题困难得多。由于轴对称问题存在流函数，对绕球流动采用柱坐标系 (r, φ, z)。取球心为坐标原点，z 轴经过球心且与主流平行，r 为垂直于 z 轴的径向坐标。因为流动是关于 z 轴对称的，因此流动参量与 φ 无关。因而可采用平面势流的处理方法近似地进行分析。

9.5.1 理想流体的绕球流动

绕球流动与空间点源、点汇及偶极流流动相关。对强度为 Q 的空间点源，在任意半径为 R 的球面上有

空间点源流动的速度势为

对于点汇流动，其势函数为

The velocity potential of the combination of a point source flow and a point sink flow [Fig. 9.16(a)], which have the same intensity of Q, yields

$$\varphi = -\frac{Q}{4\pi}\left(\frac{1}{R_1} - \frac{1}{R_2}\right) \quad (9.24)$$

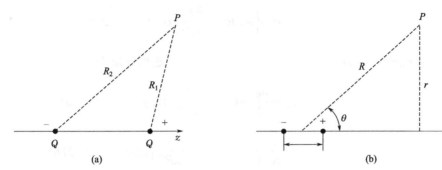

Fig. 9.16 Spatial doublet flow
图 9.16 空间偶极流

where R_1 and R_2 are the distances from any point P to the two sources respectively. As shown in Fig. 9.16(b), a spatial doublet flow can be created by making the point source and point sink be infinitely close, and with the decreasing distance dz, the intensity Q is increased to ensure the dipole moment M ($M = Qdz$) to be a limited value. Its potential function is

$$\varphi = -\frac{Q}{4\pi}d\left(-\frac{1}{R}\right) = \frac{Qdz}{4\pi}\frac{d\left(\frac{1}{R}\right)}{dz}$$

Because $R = \sqrt{r^2 + z^2}$

$$\frac{d\left(\frac{1}{R}\right)}{dz} = -\frac{1}{R^2}\frac{dR}{dz} = -\frac{z}{R^3} = -\frac{1}{R^2}\cos\theta$$

Then the velocity potential of spatial doublet flow is obtained as

$$\varphi = -\frac{M}{4\pi R^2}\cos\theta \quad (9.25)$$

Similar to a planar potential flow around a cylinder, the flow around a sphere can be regarded as the superposition of a uniform flow paralleling to the z-axis with the velocity of v_∞ and a doublet flow along the z-axis. So the velocity potential is

的偶极流的叠加，因此速度势为

$$\varphi = \varphi_1 + \varphi_2 = v_\infty z - \frac{M}{4\pi R^2}\cos\theta$$

i. e.

即

$$\varphi = \left(v_\infty R - \frac{M}{4\pi R^2}\right)\cos\theta \quad (9.26)$$

The velocity components are

速度分量为

$$v_r = \frac{\partial \varphi}{\partial R} = \left(v_\infty + \frac{M}{2\pi R^3}\right)\cos\theta \quad (9.27a)$$

$$v_\theta = \frac{1}{R}\frac{\partial \varphi}{\partial \theta} = -\left(v_\infty - \frac{M}{4\pi R^3}\right)\sin\theta \quad (9.27b)$$

On the spherical surface where $R = R_0$, $v_r = 0$, so Eq. (9.27c) can be obtained according to Eq. (9.27a)

在球壁面 $R = R_0$ 处，径向分速度 $v_r = 0$，由式(9.27a) 可得

$$M = -2\pi v_\infty R_0^3 \quad (9.27c)$$

Substituting Eq. (9.27c) into Eq. (9.26) gives the potential function as

将式(9.27c) 代入式(9.26)，得到绕球流动的势函数为

$$\varphi = v_\infty R\left[1 + \frac{1}{2}\left(\frac{R_0}{R}\right)^3\right]\cos\theta = v_\infty z\left[1 + \frac{1}{2}\left(\frac{R_0}{R}\right)^3\right] \quad (9.28)$$

According to Eq. (8.12) and Eq. (9.28), the stream function is obtained.

由式(8.12)、式(9.28) 得到流函数为

$$\psi = \frac{1}{2}v_\infty R^2\left[1 - \left(\frac{R_0}{R}\right)^3\right]\sin^2\theta \quad (9.29)$$

Equation (9.29) indicates that $\psi = 0$ on both the sphere surface ($R = R_0$) and the entire z-axis. The velocity on the sphere, v_θ, can be obtained from Eq. (9.27b) and Eq. (9.27c).

上式表明，在 $R = R_0$ 的球面及整个 z 轴上 $\psi = 0$。由式(9.27b) 和式(9.27c) 可得球面上的速度为

$$v_\theta = -\frac{3}{2}v_\infty \sin\theta \quad (9.30)$$

The pressure distribution on the sphere surface is

球面上的压力分布为

$$p_s = p_\infty + \frac{\rho}{2}(v_\infty^2 - v_\theta^2)$$

The dimensionless pressure is in the form

表示成无因次压力形式

$$\bar{p}_s = \frac{p_s - p_\infty}{\frac{\rho}{2}v_\infty^2} = 1 - \left(\frac{v_\theta}{v_\infty}\right)^2 = 1 - \frac{9}{4}\sin^2\theta \qquad (9.31)$$

9.5.2 Viscous Fluid Flow Around a Sphere

Viscous fluid flow around a sphere is very complicated due to the formation and separation of boundary layer. Some conclusions about it are resulted from experimental investigations.

Figure 9.17 illustrates the experimental pressure distributions on sphere surface corresponding to different Re_d. In the range of $\alpha < 45°$, the pressure distribution of viscous fluid flow is very similar to that of ideal fluid flow expressed by Eq. (9.31). As α increases, the pressure is affected by Re_d and boundary layer properties. When Re_d is very small ($Re_d < 10$), there is no obvious boundary layer separation, and the flow is very similar to an ideal fluid flow around a sphere. As Re_d increases, the boundary layer separates from the sphere at point S (Fig. 9.18). As can be seen from Fig. 9.17 and Fig. 9.18, the boundary layer separation point S locates at the downstream in terms of point P where the pressure is the lowest, and usually point S locates at the point of $\alpha \approx 83°$.

Once the laminar boundary layer separates from the boundary, it will be taken away by free stream. At point T, the unsteady laminar boundary layer flow transmits to be steady turbulent boundary layer flow. As Re_d increases further, point T moves back toward the sphere surface and may coincide with point S. Then the boundary layer near point S becomes a turbulent flow. Since the fluid in turbulent boundary layer has stronger mixing effect than that in laminar boundary layer, it can get more kinematic energy from free stream to make up the frictional energy loss. As a result, point S moves down the stream.

9.5.2 黏性流体的绕球流动

由于存在黏性边界层及其绕流边界层的脱离，实际黏性流体绕流圆球的情况相当复杂，现有研究结果来自于试验研究。

图9.17为不同 Re_d 下实验得到的球面压力分布曲线。当 $\alpha < 45°$ 时，实际流体与式(9.31)表示的理想流体绕流的压力分布十分接近。随着 α 的增大，压力分布开始受 Re_d 和边界层性质的影响。当 Re_d 很小（$Re_d < 10$）时，没有明显的边界层分离，流动非常接近于理想流体的绕流。当 Re_d 逐渐增大时，边界层在 S 点处与球面分离，如图9.18所示。对照图9.17及图9.18可看出，分离点通常在最低压力点 P 的下游，一般位于 $\alpha \approx 83°$ 处。

在 S 点上分离的层流边界层被主流引射走后，很不稳定，到 T 点处就转变为比较稳定的湍流状态。随着 Re_d 进一步增大，位于球体外的 T 点逐渐向球面靠近，最后可与脱离点 S 重合，S 点附近的边界层从而就变成湍流边界层。湍流边界层中流体的混合作用比层流边界层强烈，从外侧主流获得能量的能力较强，于是脱离点 S 开始向下游（球后部）移动。

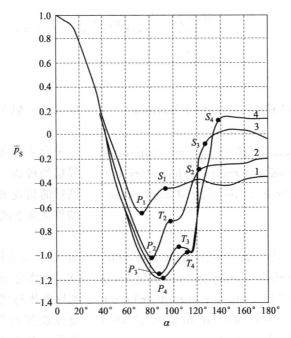

Fig. 9.17 Pressure distribution on sphere surface
图 9.17 球面上的压力分布

1—$Re_d=157200$; 2—$Re_d=251300$; 3—$Re_d=298500$; 4—$Re_d=424500$

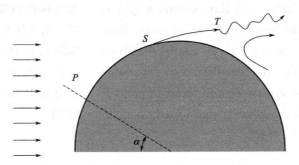

Fig. 9.18 Boundary layer separation of viscous fluid flow around sphere
图 9.18 黏性流体绕球流动的边界层分离

P—the lowest pressure point 最低压力点；S—boundary layer separation point 边界层分离点；
T—transition point from laminar boundary flow to turbulent boundary flow 层流与湍流的转戾点

It was revealed that when Re_d is larger than a critical value, point S moves from $\alpha\approx83°$ to $\alpha\approx140°$, the flow status is optimized.

研究发现，Re_d 超过临界值，分离点 S 由 $\alpha\approx83°$ 处逐步移至 $\alpha\approx140°$ 处，绕流状况得到改善。

Exercises

9.1 In fluid mechanics, how does a boundary layer form?

习题

9.1 流体力学中的边界层是如何形成的？

9.2 Describe the definitions of displacement and momentum thicknesses.

9.3 What are the differences between the flows of an ideal fluid and a viscous fluid around a cylinder?

9.4 Why does the boundary layer separation occur?

9.5 What is the contribution of the boundary layer theory to the development of fluid mechanics?

9.6 In the boundary layer sticking to a smooth plate, the shear stress at a point on the wall, τ_w, is related to the fluid density, ρ, dynamic viscosity, μ, the distance from the point to the leading edge of the plate, l, and the bulk flow velocity, v_∞. Apply the Π theorems to establish a relationship between the shear stress, τ_w, and the four influencing factors.

9.7 Explain the formation mechanism of the Kármán vortex street.

9.8 Determine the shedding off frequency and Kármán force of the Kármán vortexes generated behind a wind-blown electric line with the diameter of 3mm. The wind velocity is 20m/s and the temperature is 273K.

Reference
参 考 文 献

1. Munson B R, Young D F, Okiishi T H. Fundamentals of fluid mechanics -4th edition. John Wiley & Sons, Inc, 2009.
2. Kundu P K, Cohen I M. Fluid Mechanics-Fourth Edition. Academic Press is an imprint of Elsevier, 2008.
3. 顾伯勤. 流体力学. 北京：中国科学文化出版社, 2002.
4. 黄卫星. 工程流体力学. 第2版. 北京：化学工业出版社, 2008.
5. Douglas J F, Gasiorek J M, and Swaffield J A. Fluid mechanics. Harlow, Essex, England Longman Scientific & Technical, 1995.
6. Joseph A. Fundamentals of fluid mechanics. New York：Wiley, 1999.
7. Kay J M, Nedderman R M. Fluid mechanics and transfer processes. Cambridge [Cambridgeshire]：Cambridge University Press, 1985.
8. Kumar D S. Fluid mechanics & Fluid Power Engineering. Ludhiana：Dewan Sushil Kumar Kataria, 1982.
9. Mathur M L, Sharma R S, Purotohit R K. Fluid mechanics & Fluid Machines. New York：Tech India Publications, 1987.
10. Noel de Nevers. Fluid mechanics for chemical engineers. New York：McGraw-Hill, 1991.
11. Sachdev P L, Venkatachalappa M. Recent advances in fluid mechanics. Amsterdam, Netherlands：Gordon & Breach Science Publishers, 1998.
12. Kraut G P. Fluid Mechanics for Technicians. New York：Macmillan Publishing Company, 1992.
13. Schlichting H. Boundary Layer Theory. New York：McGraw-Hill Book Company, 1979.
14. White F M. Fluid Mechanics. New York：McGraw-Hill Book Company, 1986.
15. 白铭声. 流体力学及流体机械. 北京：煤炭工业出版社, 1980.
16. 费祥麟. 高等流体力学. 西安：西安交通大学出版社, 1989.
17. 江宏俊. 流体力学. 北京：高等教育出版社, 1985.
18. 李翼祺, 马素贞. 流体力学基础, 北京：科学出版社, 1983.
19. 廖艾贤. 流体力学. 北京：中国铁道出版社, 1987.
20. 刘光宗. 流体力学原理与计算基础. 西安：西安交通大学出版社, 1992.
21. 刘天宝. 流体力学及叶栅理论. 北京：机械工业出版社, 1983.
22. 罗大海, 诸葛茜. 流体力学简明教程. 北京：高等教育出版社, 1986.
23. 牟乃让. 流体力学与传热学基础. 北京：机械工业出版社, 1984.
24. 潘文全. 工程流体力学. 北京：清华大学出版社, 1988.
25. 王维新. 流体力学. 北京：煤炭工业出版社, 1986.
26. 王致清. 流体力学基础. 北京：高等教育出版社, 1987.
27. 叶敬棠. 流体力学. 上海：复旦大学出版社, 1989.
28. 张也影. 流体力学. 北京：高等教育出版社, 1998.
29. 张兆顺, 崔桂香. 流体力学. 北京：清华大学出版社, 1999.
30. 郑洽馀, 鲁钟琪. 流体力学. 北京：机械工业出版社, 1980.
31. 周谟仁. 流体力学泵与风机. 北京：中国建筑工业出版社, 1985.
32. 庄礼贤. 流体力学. 合肥：中国科学技术大学出版社, 1991.
33. 沈钧涛, 鲍慧芸. 流体力学习题集. 北京：北京大学出版社, 1990.
34. 张也影, 王秉哲. 流体力学题解. 北京：北京理工大学出版社, 1996.
35. 朱之墀, 王希麟. 流体力学理论例题与习题. 北京：清华大学出版社, 1986.